Farmers and Agriculture in the Roman Economy

W0018421

Often viewed as self-sufficient, Roman farmers actually depended on markets to supply them with a wide range of goods and services, from metal tools to medical expertise. However, the nature, extent, and implications of their market interactions remain unclear. This monograph uses literary and archaeological evidence to examine how farmers – from smallholders to the owners of large estates – bought and sold, lent and borrowed, and cooperated as well as competed in the Roman economy. A clearer picture of the relationship between farmers and markets allows us to gauge their collective impact on, and exposure to, macroeconomic phenomena such as monetization and changes in the level and nature of demand for goods and labor. After considering the demographic and environmental context of Italian agriculture, the author explores three interrelated questions: what goods and services did farmers purchase; how did farmers acquire the money with which to make those purchases; and what factors drove farmers' economic decisions? This book provides a portrait of the economic world of the Roman farmer in late Republican and early Imperial Italy.

David B. Hollander earned a PhD in Ancient History from Columbia University (2002). He published his monograph *Money in the Late Roman Republic* in 2007 and served as the economy editor for *The Encyclopedia of Ancient History* (2012). He is an associate professor in the History Department at Iowa State University, USA.

Cover image: Denarius of C. Marius Capito. Photo by author.

Farmers and Agriculture in the Roman Economy

David B. Hollander

Routledge
Taylor & Francis Group

LONDON AND NEW YORK

First published 2019
by Routledge
2 Park Square, Milton Park, Abingdon, Oxon OX14 4RN

and by Routledge
52 Vanderbilt Avenue, New York, NY 10017

First issued in paperback 2020

Routledge is an imprint of the Taylor & Francis Group, an informa business

British Library Cataloguing-in-Publication Data
A catalogue record for this book is available from the British Library

Library of Congress Cataloguing-in-Publication Data
A catalog record for this book has been requested

ISBN 13: 978-0-367-66622-4 (pbk)
ISBN 13: 978-1-138-09988-3 (hbk)

Typeset in Times New Roman
by codeMantra

Contents

Acknowledgements   ix

1   Problems and sources   1
   1.1   Introduction 1
   1.2   The problem with self-sufficiency 1
   1.3   Sources for Roman agriculture 4
      1.3.1 Cato's *De agricultura* 4
      1.3.2 Varro's *De re rustica* 6
      1.3.3 Columella's *De re rustica* 7
      1.3.4 Pliny the Elder's *Natural History* 7
      1.3.5 Archaeology 10
      1.3.6 Art 11
   1.4   Conclusion 15

2   The parameters of Roman agriculture   20
   2.1   Introduction 20
   2.2   Climate and geography 20
   2.3   Demography 22
   2.4   Roman crops 23
      2.4.1 Cereals 23
      2.4.2 Viticulture 27
      2.4.3 Olives 28
      2.4.4 Other fruit trees 29
      2.4.5 Legumes 29
      2.4.6 Fiber crops 30
      2.4.7 Nuts 31
      2.4.8 Vegetables and tubers 32
      2.4.9 Other plants 33
   2.5   Livestock 33
   2.6   Conclusion 35

3   Buyers and borrowers: The rural demand for goods, services, and money   40

3.1   *Introduction 40*
3.2   *Start-up requirements 41*
        3.2.1 Building supplies 42
        3.2.2 Metal agricultural tools 43
        3.2.3 Wooden tools 46
        3.2.4 Livestock 46
        3.2.5 Storage 46
        3.2.6 Clothing 47
        3.2.7 Processing equipment 47
        3.2.8 Other equipment 48
3.3   *Seasonal requirements 48*
3.4   *Maintenance requirements 49*
3.5   *Extra-agricultural expenses 51*
3.6   *Borrowing and sharing 52*
3.7   *Rural demand for coinage 53*
3.8   *Conclusion 57*

4   Vendors and lenders: The rural supply of goods and services   62

4.1   *Introduction 62*
4.2   *Animals and animal byproducts 63*
        4.2.1 Meat 64
        4.2.2 Wool 66
        4.2.3 Dairy products 67
        4.2.4 Apiculture 68
4.3   *The profitability of plants 71*
        4.3.1 Grain 71
        4.3.2 Viticulture 72
        4.3.3 Olive oil 73
        4.3.4 Flowers 73
        4.3.5 Vegetables 73
        4.3.6 Other produce 74
        4.3.7 Linen 75
4.4   *The sale of the superfluous 75*
4.5   *Working for others 76*
4.6   *Moneylending 77*
4.7   *Conclusion 78*

5   Farmers' markets, farmers' networks   83

5.1 *Introduction 83*
5.2 *Markets 85*

5.3 Reciprocity 87
5.4 Redistribution 89

6   Farmers in Roman economic history   93
6.1   Introduction 93
6.2   Degrees of market dependency 94
     6.2.1 Elite farmers 94
     6.2.2 Moderately wealthy farmers 95
     6.2.3 Smallholders 95
     6.2.4 Landless farmers 96
6.3   Farmers in the Roman economy 96
     6.3.1 The second century BCE 97
     6.3.2 The first century BCE prior to the reign of Augustus 99
     6.3.3 The early Empire 100

Bibliography   105
Index   125

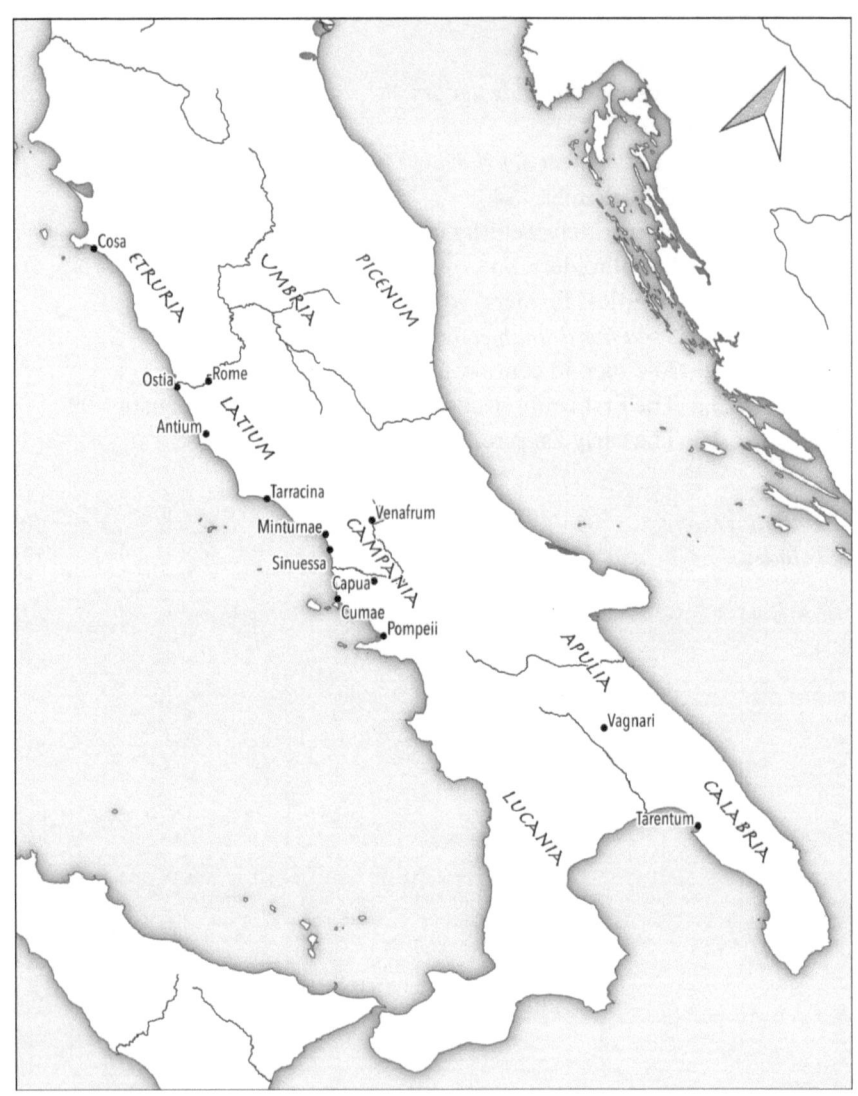

Map of Italy made using QGIS. Ancient World Mapping Center. "Ba_rivers", "Openwater" and "Coastline". Accessed February 9, 2018. http://awmc.unc.edu/wordpress/map-files/

Acknowledgements

I first became interested in the economic behavior of Roman farmers while in graduate school; my initial thoughts on this topic can be found in my dissertation and the subsequent book, *Money in the Late Roman Republic* (2007). At Iowa State University, with its long commitment to the study of all things rural and agricultural, it seemed like an obvious topic to pursue further. Hence thanks must go first and foremost to my colleagues past and present at Iowa State, beginning with a series of supportive department chairs: Andrejs Plakans, Charles Dobbs, Pamela Riney-Kehrberg, Michael Bailey, and currently Simon Cordery. Many colleagues and students in what is now the RATE program have helped me see agricultural issues from new perspectives. Grants from the University, College of Liberal Arts and Sciences, and the Center for Excellence in the Arts and Humanities facilitated my research and writing. I am also grateful to my classical colleagues in the department of World Languages and Cultures, especially Peggy Mook and Rachel Meyers. Larry Elsken and Donna Gatewood fielded my occasional questions about veterinary medicine with good humor and Donna, along with Becky Olson, provided some insight into cheese-making. I also appreciate the help of the staff at the University Library, especially those in Interlibrary Loan.

Preliminary versions of some parts of this book were presented at conferences and elsewhere over the years; I therefore thank session organizers and audiences at annual meetings of the Archaeological Institute of America, Classical Association of the Middle West and South, Society of Biblical Literature, and Agricultural History Society, as well as hosts and audiences at the Institute for the Study of the Ancient World at New York University, the University of Leicester, and the University of Missouri. Thanks also to Ethan Spanier who provided some early feedback on the project. I am especially grateful to the anonymous reviewers of my initial book proposal and the subsequent manuscript for their many helpful comments. Of course, none of them are responsible for any remaining flaws. I also greatly appreciate the help (and patience) of my editor at Routledge, Michael Greenwood. The most patient, however, must surely be my wife and daughter, to whom I dedicate this work.

1 Problems and sources

1.1 Introduction

The vast majority of the population of Roman Italy worked the land,[1] but collectively Roman farmers have not received the attention from economic historians that has, in recent years, been devoted to fullers, prostitutes, traders, shopkeepers, auctioneers, and bankers.[2] This is not to say that they have been ignored – there has been considerable interest in certain subsets of Roman farmers such as tenants or estate owners[3] – but they deserve greater attention as a whole. This book examines not Roman farming *per se*, a topic on which quite a lot has been published, but rather the economic behavior of Roman farmers: what they bought and sold, how they employed labor and capital, as well as their impact on and relation to the broader Roman economy.

For the purposes of this study, I define 'farmer' broadly as someone engaged in agriculture on a full or part-time basis, including laborers as well as those working in a management capacity either on their own land or someone else's property. Thus, farmers can be the members of a smallholder household, seasonal workers from a nearby village or town (or even further afield), an estate owner, or his *vilicus* (estate manager). The chronological and geographical confines are late Republican and early imperial Italy (200 BCE to 200 CE), arguably the best documented region and period. Although I will make occasional references to other times and places, a broader scope would have made the project unmanageable.

1.2 The problem with self-sufficiency

One of the most persistent ideas about Roman farmers is that they were largely self-sufficient and so had a limited impact on the broader Roman economy. Scholars generally agree that the Roman elite and especially the agricultural writers considered self-sufficiency or autarky a "moral precept" and the "prevailing ideology."[4] While there is increasing doubt about the possibility of *achieving* self-sufficiency (see below), many historians consider it to have been at least the goal of most Roman peasants.[5] The pursuit of self-sufficiency, it is often further believed, insulated rural areas from the monetized economy. As Keith Hopkins once put it, the "solid mass of self-sufficient production always stood outside the money economy."[6]

At first glance, it is easy to see why self-sufficiency remains a powerful idea in discussions of Roman farmers. In the second century BCE Cato the Elder declared that the farmer should be a seller, not a buyer (*Agr.* 2.7); Varro, in a dialogue written a little more than a century later, has a character recommending that "nothing should be purchased which can grow on the farm and be made by the household" (*Rust.* 1.22); and the early imperial encyclopedist Pliny the Elder said that it was an old and well-accepted adage that "whoever buys what his farm could give him is a worthless farmer" (*HN* 18.41).[7]

Given such statements, it is hardly surprising that the concept of self-sufficiency, comes up so frequently in modern discussions of Roman farming. However, the matter is far more complicated and problematic than it would appear. Most importantly, *all* of the ancient agricultural writers make it clear in other, perhaps less memorable passages, that there were plenty of things a farmer *had* to purchase. Since that is a major theme of this study, I will leave the issue of necessary purchases to one side for the moment and focus on the idea of self-sufficiency in modern scholarship about Roman farmers. First of all, while scholars regularly use the term 'self-sufficiency,' they have not done so in a consistent fashion. Indeed, in many instances they have used the term without any indication of what they mean by it.[8] Some conceive of self-sufficiency as an ideal rather than an economic practice that farmers could actually achieve. Paul Erdkamp, for example, calls autarky an "important goal" but concedes that "pure autarky could not be achieved."[9] Some consider self-sufficiency to be a phenomenon happening at the level of the individual household, while others see it as a village, town, or regional phenomenon.[10] Some, when they talk about self-sufficiency, just mean self-sufficiency in all the *food* consumed by a farm household rather than in other goods like tools, ceramics, clothing, and so forth,[11] and acknowledge that even peasants had to make purchases in the market.[12] Sometimes self-sufficiency is part of a strategy that involves specialized production *for* the market.[13]

Although some scholars have viewed the pursuit of self-sufficiency as a dangerous strategy,[14] others have regarded it as "hard-hearted economic rationality,"[15] a "necessity,"[16] or "a rational response to conditions of risk and uncertainty."[17] A few scholars have even pointed to the institution of the *nundinae* (the markets held every eighth day) as proof for some kind of self-sufficiency.[18] Joan Frayn argued that the *nundinae* "by their very name suggest that they were intended to supplement the living of a farming community which was largely self-sufficient."[19] But surely the best proof would be the *absence* of markets or at least a considerably longer interval between market days. A weekly market day strongly suggests that there was a regular demand for goods that could not be supplied from one's own farm or estate.[20] Poor transportation technology and networks have also been cited as the cause of Roman self-sufficiency. J. K. Evans, for example, argued that "the inadequacy of Roman land transport effectively condemned [the] majority to self-sufficiency."[21]

There has, nevertheless, been plenty of skepticism about the idea. As early as 1970 Martin Frederiksen noted that there was "something a little suspect in the ideal of self-sufficiency."[22] In 1979, Paul Veyne published an article entitled "Mythe et réalité de l'autarcie à Rome" in which he discussed "une curieuse contradiction: que les Anciens (et leurs historiens) en aient tellement parlé et qu'ils l'aient si peu practiquée."[23] Subsequently, many scholars began to refer to self-sufficiency as a myth[24] or a mirage.[25] In *The Corrupting Sea*, Peregrine Horden and Nicholas Purcell observed that "the prevalence of autarky has been deduced from its persistence as an ideal: practice has been inferred from rhetoric."[26] They call "the pursuit of self-sufficiency ... an ethical tenet rather than a practice observable in reality."[27] More recently, in her excellent book *Shopping in Ancient Rome,* Claire Holleran pointed out that "the very fact that self-sufficiency was something to boast about suggests that this was not universal practice."[28] Despite these doubts, historians continue to use the term but, with no generally accepted definition of self-sufficiency and an emerging consensus that, even if it was an ideal, it was not an achievable one, it is probably best to forgo the use of the term 'self-sufficiency' altogether and instead employ less ambiguous terminology such as 'degrees of market dependency.' A major goal of this book is to discuss what degrees of market dependency are plausible for various sorts of Roman farmers. Furthermore, I want to question some core assumptions of the modern discourse on self-sufficiency. Are our sources really advocating self-sufficiency as consistently and unambiguously as we tend to assume? It is worth noting that the Roman agricultural writers never use the word *autarkeia* nor any other term denoting the concept of self-sufficiency. Even if the pursuit of self-sufficiency (in some form) is in fact a rational economic strategy, it would be foolish to assume Roman farmers always behaved rationally.[29] The field of behavioral economics has identified many situations in which humans make economic decisions that are neither optimal nor unbiased.[30] I would suggest that many Roman farmers were *not* especially rational with respect to maximizing profit and that there were many different ideas about the best way to manage a farm.[31] As the discussion of sources (below) indicates, advocating self-sufficient strategies could simply be a convenient way to attack one's flashier, more market-reliant political rivals. Furthermore, was self-sufficiency (i.e., a very low degree of market dependence) a realistic goal from the perspective of resources and labor? In other words, could a Roman get anywhere close to self-sufficiency without vast estates to provide all the raw materials for his or her food, shelter, and clothing, and an army of slaves and/or dependents to farm the land, care for the animals, spin, weave, and make tools to provide even the *minimum* requirements of survival? Petronius' Trimalchio seems to approach those conditions (*Sat.* 38) but surely no real Roman could aside from an emperor. Part of the appeal, finally, of the idea of self-sufficiency is that diminishing one's reliance on the market provides security. But there were two far more practical ways to achieve security: friends and cash.[32] Sensible Romans sought out both and, of course, neither could be acquired by pursuing self-sufficiency. This brings me to one further goal of this study: to consider not just the rural demand for goods and services but also rural demand for money.[33]

1.3 Sources for Roman agriculture

This study draws on a wide range of textual and archaeological sources about which some preliminary remarks are in order. Four literary works are of particular importance: Cato's *De agricultura*, Varro's *De re rustica*, Columella's *De re rustica*, and Pliny the Elder's *Naturalis historia*. Due to their prominence in the following chapters, it is worthwhile discussing their value (and the problems associated with their use) as evidence immediately.

1.3.1 Cato's De agricultura

Marcus Porcius Cato or Cato the Elder (234–149 BCE) was a prominent Roman politician, orator, and writer in the first half of the second century BCE. As a young man he fought against Hannibal in the Second Punic War and, in his final years, he helped bring about the Third Punic War and the destruction of Carthage. According to Plutarch (*Cat. Mai.* 3, 25), he was always devoted to agriculture, working beside his slaves in his youth and writing, probably in the 150s,[34] the *De agricultura,* his only surviving work. After briefly praising farmers as being the strongest men and most vigorous soldiers, Cato proceeds to discuss a range of subjects including how to go about purchasing a farm, what crops should be planted in what kinds of soil, the price of mills, and who makes the best press-ropes. Towards the end of the work Cato praises cabbage (*brassica*) at considerable length (*Agr.* 156–8)[35] and shares a spell to cure a sprain (*Agr.* 160). Even in antiquity this work seems to have been viewed as somewhat odd. Varro and Plutarch both found the inclusion of various recipes rather strange (Plut., *Cat. Mai.* 25; Varro, *Rust.* 1.2.28). Nevertheless, later Roman agricultural writers clearly respected some of Cato's opinions, though they did not accept them uncritically. While those who study Roman farming have no choice but to rely heavily on Cato, in recent decades scholars have grown increasingly skeptical about the *De agricultura* as a straightforward source for agricultural practices. In 1970 K. D. White acknowledged the "rambling style and incoherent organization" of the work, but defended it as having an "abundance of shrewd common sense and practical farming knowledge."[36] Subsequent assessments have been less kind. Steven Spurr called it "an incomplete and random assemblage of agricultural precepts,"[37] while John Bodel contrasts the high literary quality of the preface with "the disjointed meandering of the main body of the work."[38] Nicola Terrenato has described it as "an apparently inexplicable combination of outdated folklore and agricultural science fiction [which] would be completely unhelpful, if not actually harmful, for a landowner newly embarking in agriculture."[39] He also criticizes the work for, on the topic of viticulture, providing "terrible advice" which "makes no economic sense" and notes that the author "seems blissfully unaware" of the existence of Greco-Italian wine amphoras despite the fact that he "goes to considerable pains to list every lid or chamber pot that one might possibly need on a farm."[40] There has also been the problem of finding actual examples of the sort of villa Cato describes in the archaeological record of Republican Italy.[41]

A more sophisticated understanding of the *De agricultura* has begun to emerge but it is one that makes it harder for those interested in the ancient economy and Roman farming *per se* to make use of Cato's evidence. According to this new view, Cato's work forms part of his broader response to cultural and economic change in the second century, most memorably illustrated by his efforts to combat luxury while censor (Plut., *Cat. Mai.* 18) and his reaction to the high-powered rhetorical ability demonstrated by the Athenian philosophical delegation of 155 BCE (Plut., *Cat. Mai.* 22). Bodel, for example, suggests that Cato, in the *De agricultura*, wanted "to reaffirm the traditional belief of the ancestors that farming produced the best citizen-soldiers."[42] Brendon Reay argues that the work should be understood in the context of aristocratic performance and self-representation: Cato "invents himself as the one expert above all others and thereby stakes out for himself a privileged position above and beyond the rest of the aristocratic pack."[43] Terrenato proposes that Cato used the descriptions of rustic simplicity in the work to contrast himself with the "vaunted *villae expolitissimae* of the hated Scipios," his political enemies.[44]

The upshot of all this is that we must be very suspicious of Cato when he makes his recommendations about how to run a farm. It is likely that some of the strategies he describes stand in stark contrast to those of many of his contemporaries. Nevertheless, his work still contains much valuable information for the economic historian. Even if the text is "prescriptive rather than descriptive,"[45] the details must be largely accurate or Cato would appear incompetent rather than an expert. So while we – along with Varro (*Rust.* 1.18) – may view with skepticism, for example, Cato's recommendation that an estate owner should have 16 slaves to operate a 100 *iugera* (approximately 62 acres) vineyard, it is quite likely that a farmer who produced wine would need much of the equipment (jars, presses, strainers, pruning tools, baskets, etc.) that Cato proceeds to list (*Agr.* 11). This study, very much concerned with the economic implications of the tools necessary to perform Roman agriculture, will pay close attention to such details, which can, of course, in some cases be backed up by archaeological evidence.

1.3.2 Varro's **De re rustica**

Marcus Terrentius Varro (116–27 BCE) was a late Republican politician who reached the office of praetor and managed to survive both having been a supporter of Pompey in the civil war with Caesar (Caes., *B Civ.* 2.20) and subsequent proscription by Marcus Antonius (Gell., *NA* 3.10.17). Of his voluminous writings, only parts of his twenty-five-volume *De lingua Latina* and the whole of his *De re rustica* survive. The latter work, composed in the 30s BCE, consists of three books, each a dialogue about some aspects of agriculture. The first book takes place at the temple of Tellus in Rome during a festival. Varro's interlocutors include his father-in-law, Gaius Fundanius, the *eques* Gaius Agrius, and the *publicanus* Publius Agrasius. These men, whose names all allude to agriculture (*fundus* means farm; *ager* means field),[46] proceed to have a general discussion of agriculture, with a focus on arable farming, while waiting for the arrival of the priest Lucius Fundilius who has invited them for dinner. Gaius Licinius Stolo ("shoot" or "sucker") and Gnaeus Tremelius Scrofa ("sow") soon join them. The dialogue ends with the shocking news, delivered by a weeping freedman, that their would-be host has been knifed to death in a crowd. The second book takes place in Epirus a few decades earlier and focuses on raising cattle. This dialogue also ends with the arrival of a freedman but this time the message is merely that Varro and Scrofa should come to the house of Vitulus ("bull-calf") right away. The final book has the Villa Publica in Rome as its setting and turns to the topic of *pastio villatica*, the raising of birds, fish, and other animals as well as beekeeping. The speakers, many of whom have avian names such as Marcus Petronius Passer ("sparrow") and Minucius Pica ("magpie"), pass the time in conversation while awaiting the results of an election. The books ends with them escorting their victorious candidate home. Varro explains at the beginning of the first book that he is writing for his wife, Fundania, who has just bought a farm, to teach her about agriculture, but it is not clear how seriously we are meant to take this claim. Although White called the work "immensely superior to that of Cato in every respect … [and] based throughout on practical knowledge and tried experiment,"[47] there are a number of curious features. Leah Kronenberg has made a strong case for seeing Varro's *De re rustica* as "a subversive work, which uses farming as a vehicle to expose the hypocrisy and pretensions of Roman morality, intellectual culture, and politics in the Late Republic."[48] This interpretation accords well with the profusion of punning names as well as the errors made by some of the speakers. Even White acknowledged that "Varro makes mistakes, some of them astonishing,"[49] but later ancient agricultural writers seem to take Varro seriously enough and satire, to have any sting, must reflect one's actual context. Still, the use of the *De re rustica* for the study of Roman farming and the ancient economy requires caution. For one thing, as Peter Brunt observed, the work has little to say about anyone other than wealthy estate owners,[50] so it would be dangerous to extrapolate from his descriptions of their practices.

1.3.3 *Columella's* De re rustica

With Lucius Iunius Moderatus Columella, we arrive at the best of the Roman agricultural writers. Columella (ca. 4–70 CE), originally from Gades, settled in central Italy following a military career and owned at least four farms (three in Latium, one perhaps in Etruria). White called him "the most accomplished stylist of the Roman agricultural writers," emphasizing his "practical knowledge" and noting that he "is the only surviving agronomist who has attempted to quantify the problem of manpower requirements by furnishing information on output measured in man-days per *iugerum*."[51] Columella's *De re rustica,* dedicated to one Publius Silvinus, consists of twelve books. Book one discusses the basic elements of the farm, such as water supply, buildings, tenants, and slaves. The second book treats arable agriculture, including plowing, sowing, manuring, harvesting, threshing, and winnowing. Books three through five mainly concern viticulture but with some discussion of olive and fruit tree cultivation towards the end. Columella turns to livestock in books six through nine, considering cattle, mules, horses, donkeys, sheep, goats, pigs, dogs, birds, and fish as well as wild animals such as deer and bees. He provides instructions on everything from training oxen to making cheese and even gives advice on how to name dogs (7.12.13). Switching to verse for book ten, Columella discusses gardens. Book eleven, in prose again, focuses on the *vilicus* and provides an agricultural calendar. The final book, on the *vilica* (wife of the *vilicus*), discusses the preservation and storage of farm produce at great length.[52] Katharine von Stackelberg calls Columella's books "by far the most comprehensive and user friendly work on ancient agriculture."[53] However, it is important to keep in mind that Columella is writing for the owners of large estates[54] and, as Arnaldo Marcone puts it, has a "clear propagandistic intent."[55] Columella proclaims agriculture to be the profession most morally acceptable for respectable Romans (*Rust.* 1.pr.10). Nevertheless, the richly detailed account of a wide range of farm practices and problems makes the *De re rustica* an extremely valuable source for the economics of Roman agriculture. In addition to information about tools and markets, Columella occasionally makes reference to the activities of less well-off farmers. For example, in his discussion of the drying of apples, he notes that dried apples sometimes make up a large share of the food of *rustici* (peasants) during the winter (*Rust.* 12.14).

1.3.4 *Pliny the Elder's* Natural History

Pliny the Elder's *Natural History* contains a considerable amount of information about Roman agriculture but, like our other sources, must be used with caution. History's most famous volcano victim was born around 23 or 24 CE at Novum Comum in northern Italy to an equestrian family. He served in the cavalry in Upper Germany, where he fought against the Chauci under the command of Gnaeus Domitius Corbulo in the late 40s, and in Lower Germany alongside the future emperor Titus. While serving as a cavalry commander, Pliny began his prolific writing career, producing a book on throwing javelins from horseback. His other works included a two volume biography of Publius Pomponius Secundus (an ex-consul and tragic poet who, Pliny tells us, never belched in his entire life); twenty volumes covering all the wars between Rome and the Germans, a project begun after Nero Claudius Drusus appeared to Pliny in a dream and asked him to defend his memory from oblivion; three books on rhetorical education; eight books on grammar, written during the dangerous later years of Nero's reign; and thirty-one books covering Roman history from 54 CE onwards. Unfortunately none of these works survive. Pliny also worked as a lawyer in Rome, was a close friend of the emperors Vespasian and Titus, and served as a procurator in Gallia Narbonensis, Africa, Hispania Tarraconensis, and possibly Gallia Belgica during the reign of Vespasian. In a letter to Baebius Macer, who wanted to acquire *all* of Pliny the Elder's works, Pliny the Younger explained how his uncle was able to be so prolific: he got by on very little sleep, always had books being read to him (on which he was constantly taking notes) even during dinner and while bathing (except when actually immersed in the water). He believed "there was no book so bad that some good could not be got out of it" (Plin., *Ep.* 3.5). He also habitually traveled by litter so he could continue his studies even on the streets of Rome. Pliny the Younger says his uncle criticized him for walking. At the time of his death in 79 CE during the eruption of Vesuvius, Pliny the Elder was commander of the Roman fleet at Misenum.

Given his career, one may doubt that Pliny had much firsthand experience of farming but the *Naturalis historia*, his only surviving work, is full of agricultural information. This encyclopedia consists of thirty-seven books treating cosmology, geography, anthropology, zoology, botany, medicine, and minerals. Pliny casts a wide net in his discussion of these topics. His treatment of minerals, for example, includes discussion of millstones (*HN* 36.135) and cisterns (*HN* 36.174) while the books on botany contain invaluable information not only about agriculture but also naval technology. Throughout the encyclopedia there are many strange digressions. As John Healy notes, "Pliny is clearly obsessed with the unusual and freakish."[56] The work seems to have remained quite popular from antiquity up until the eighteenth century despite fairly consistent criticism of the quality of the prose. One classicist, for example, described Pliny as someone "who can hardly frame a coherent sentence."[57] This is a bit surprising considering Pliny had also written on grammar and rhetoric – some have suggested the work was not yet revised at the time of Pliny's death.

While dedicated to the emperor Titus, Pliny wrote in his preface that the work was meant for farmers and artisans as well as students (*HN* pr.6). Probably this was simply an attempt to be self-deprecating since the text is hardly organized in a way that would make it easy for a farmer or an artisan to use. There is also the problem that he is wrong, as David Thurmond puts it, "with frustrating regularity."[58] Often excerpted (especially for art historians and those interested in ancient science and technology), in recent decades scholars have attempted to appreciate and understand the *Natural History* as a whole. Andrew Wallace-Hadrill calls Pliny a crusader and "sort of proto-environmentalist" who "by showing how nature is designed for man, [tried] to persuade man how properly to make use of his natural environment."[59] Despite his considerable and often uncritical reliance on other sources rather than firsthand experience, Pliny's encyclopedia nonetheless offers valuable information.[60]

Although Stephen Dyson has strongly argued against the ability of Cato and Varro's works to convey an "actually existing or even hoped-for agrarian reality,"[61] the *details,* if not the overall portrait, can help us understand Roman agriculture. As Geoffrey Kron has pointed out, the advice of the agronomists on, for example, game and fish farming, is "detailed and extremely well researched and sound, corresponding closely to the practice of contemporary game farmers."[62] Quirky though they might be, Cato, Varro, Pliny the Elder, and especially Columella can help us make sense of the economic life of Roman farmers. While they do not provide much direct insight into the activities of smallholders, they do indicate some of the constraints they operated under. Similarly, they do not tell us much about the marketing of produce (or the purchase of supplies), but they frequently hint at what a farmer could offer for sale or need to acquire.

A host of other textual evidence also sheds occasional light on the Roman agricultural economy, most notably the writings of Cicero, letters of Pliny the Younger, Lucretius' *De rerum natura,* and Vergil's *Georgics.* The use of bucolic poetry to understand Roman agriculture is particularly fraught with difficulties. The countryside was a popular subject in Roman poetry but works like the *Eclogues* of Virgil and Calpurnius Siculus, the Pseudo-Vergilian *Moretum,* and various Latin Priapeia do not attempt to describe the contemporary rural world in a particularly realistic manner. The *Georgics* is the most useful of such works since it is part of "the tradition of didactic poetry."[63] As Aude Doody points out, Virgil makes considerable use of Varro while Columella and Pliny the Elder later have "a serious engagement" with its content, indicating that they did not dismiss his input.[64] However, much bucolic poetry is only ostensibly about the countryside. For example, as Philip Thibodeau points out, there was "a long tradition of poems which represent an elderly man tending a garden."[65] The gardens of Simulus in the *Moretum* and the Corycian *senex* of the *Georgics* are designed in part to engage with previous examples of the genre rather than to accurately depict contemporary horticultural practice. Furthermore, a rural setting or subject could also be chosen by a poet in order to examine moral issues, as a contrast with the corrupting influence of city life, a popular theme.[66] Despite these difficulties, I will have recourse, albeit relatively infrequently, to the Roman poets' agricultural writings, especially Virgil's

Georgics. While these works do not attempt to teach agriculture in the way Cato, Varro, Columella, and Pliny do, they nonetheless offer some valuable insights.

1.3.5 Archaeology

Archaeology provides a different perspective on Roman agriculture, supplementing the literary evidence and sometimes challenging the conclusions scholars have drawn from it. A considerable number of excavations of Roman agricultural sites have been conducted throughout Italy ranging from lowly farmhouses to great villas (though it is fair to say that structures associated with the upper end of the economic spectrum have received the bulk of the attention). Kron notes that "small and medium-sized farms ... have yet to receive adequate attention."[67] Of particular importance are the excavations of sites destroyed by the eruption of Vesuvius in 79 CE. Many of the remains recovered from these sites are rare or virtually unknown from elsewhere. They include a large number of iron tools, carbonized food, as well as human and animal remains. For example, twenty-three Roman agricultural tools in a state of "remarkable preservation," now in the collections of Chicago's Field Museum and the Kelsey Museum of Archaeology in Ann Arbor, were uncovered during the excavation of a villa at Boscoreale.[68] The tools include various kinds of hoes (*rastra* and *sarcula*), hayforks (*furcae ferreae, furcillae,* and *pastina*), and axes (*secures*). Wilhelmina Jashemski and other scholars have been able to reconstruct Roman gardens with an astonishing level of detail. Her 1973–4 excavation of the Garden of the Fugitives at Pompeii, for example, was able to identify water channels and root cavities, revealing both the crops grown and details of their cultivation.[69] Settefinestre, a productive villa at Cosa excavated in the 1970s and early 1980s, has played an especially important role in attempts to reconstruct Roman farming. Perhaps the "best studied villa in the Roman world,"[70] Settefinestre may have relied heavily on slave labor and produced a considerable amount of wine for export.[71] Annalisa Marzano's *Roman Villas in Central Italy* catalogs over four hundred and fifty villas in Latium, Tuscany, and Umbria. Plenty of rural sites are known from the rest of the Italian peninsula as well.[72] An important recent development has been the excavation of sites on the lower end of the economic spectrum. The Roman Peasant Project has investigated several small, non-élite sites in Tuscany, including part of a late Republican village at Pievina and a "small-scale agro-processing point" at Case Nuove, Cinigiano.[73]

Excavations can offer a tremendous amount of information about the economics of Roman farming. Amphorae, which appear on rural and urban sites as well as in shipwrecks, provide us with excellent evidence for the long-distance distribution (via trade and other mechanisms) of wine, olive oil, and other foodstuffs.[74] Finds of coinage, non-local ceramics, and other goods provide some sense of the extent to which a particular farm was interacting with the broader economy. The size of a granary or the number of presses hint at the scale of production. Pollen samples can indicate what was grown in nearby fields.[75] The analysis of animal bones can suggest whether the animals were kept primarily for meat production, labor, or secondary products. Much can be inferred even from butchery marks. However, there are many things an excavation cannot recover. Some organic materials simply do not survive for long while valuable items often would have been stolen, salvaged, or recycled when a site was abandoned. Most importantly, perhaps, it is almost always impossible to know who owned a particular farm or how much of the surrounding land pertained to a particular farmhouse or villa.[76] Without that information we cannot identify with any precision the socio-economic status of the owner since the literary evidence makes it clear that wealthy Romans owned many farms, of varying size, often scattered around Italy. Similarly, excavations have tended to concentrate on the wealthier sites, skewing our perception of the Roman countryside in much the same way that the literary sources do.

Field survey is another useful but problematic archaeological tool. Surveys from all over the peninsula give us a sense of how the countryside was settled with the potential to inform us about population changes and how the land was exploited. However, there are problems of site identification, classification, and interpretation. Kim Bowes et al. argue that "the narrow functional categories often applied to Italian field survey data conceal a huge range of functional diversity."[77] Robert Witcher has pointed out, for example, that "loom weights and querns routinely are taken as evidence of production, despite their frequent occurrence in ritual contexts."[78] The considerable role played by finewares in the dating of sites, Alessandro Launaro argues, means that the rural habitations of the poor are "likely to be under-represented."[79] Indeed, Dominic Rathbone suggests that "most small Roman farmsteads were too flimsy and materially poor to leave much of an identifiable archaeological trace."[80] Centuriation, finally, the systematic division of land into rectangular units by surveyors (*agrimensores*) as an initial step to facilitate its distribution, offers tantalizing but limited evidence for rural settlement since, while the grids are readily visible from the air, they are often quite hard to date.[81] Despite their limitations, surveys can highlight regional differences in the development of Italian agriculture.[82] They have also made valuable contributions to the study of the Roman rural economy, calling into question, for example, the traditional explanation for the Gracchan crisis of the late second century BCE.[83]

1.3.6 Art

Roman literature strongly associated farming with a specific set of traditional values. Roman writers regularly describe farming as the most respectable occupation and the one most able to provide the state with good soldiers. The story of Cincinnatus, called from his plow to save a Roman army, is the most famous example of the noble farmer genre but the general idea that farming produced the best soldiers and most virtuous citizens is repeated *ad nauseam* over the centuries. At the very beginning of his agricultural manual, Cato the Elder states that farmers make the strongest men, most vigorous soldiers, and least mischievous citizens (*Agr.* pr.4). In 80 BCE Cicero made a similar point in his defense of Sextus Roscius, stating that "rustic life ... is the teacher of frugality, diligence, and righteousness" (*Rosc. Am.* 75). In his later *De officiis*, perhaps the definitive expression of this attitude, Cicero, after surveying other professions, declares that nothing is "better, more productive, sweeter, or more worthy of a free man than agriculture" (*Off.* 1.151). Under the early Empire we find similar remarks. Columella, in his agricultural manual, describes farming as the occupation most befitting free men (*Rust.* 1.pr.10), while Pliny the Elder refers back to Cato on this issue (*HN* 18.26).

Although Latin literature makes the association between virtue and farming abundantly clear, it is much less evident in the visual arts. The ways painters depict rural life in early imperial landscape frescos may reflect the waning interest in agriculture reported by contemporary writers. Columella, for example, bemoans the fact that he knows of neither teachers nor students of agriculture (*Rust.* 1.pr.5).

Surviving early imperial frescos frequently depict villas, gardens, and the *products* of agriculture and horticulture,[84] but, of the three staples of the ancient Mediterranean diet, grain, olive oil, and wine, only the last received much attention in Roman painting of the late Republic and early Empire.[85] Depictions of grain cultivation seem especially rare. In fact, a now lost painting from the House of Triptolemus is the only fresco I know of in which grain is a focal point rather than a minor decorative element.[86]

This curious lacuna is most evident if we turn to landscape paintings. For example, the frescos (perhaps dating to the 20s BCE) from the Villa Farnesina in Trastevere, now displayed in the Museo Nazionale Romano (Palazzo Massimo), feature scenes of cows, goats, dogs, shepherds, and possibly sheep – major elements of *pastoral* agriculture – but no plowing of fields, harvesting of grain, olives, or grapes, nothing relating to arable farming. Landscape frescos from Boscotrecase have similar features as do those from Oplontis and Pompeii. Roger Ling sums up the elements of these landscapes as follows: "trees, temples, country houses, rustic shrines, statues, worshippers, wayfarers, fishermen, goatherds, grazing flocks and cattle, [and] mere bystanders."[87] Again pastoral agriculture sometimes makes an appearance but not *arable* farming.

Of course only a small fraction of Roman frescos survive. Fortunately, two early imperial authors provide descriptions of the landscape genre, suggesting that extant examples are representative. Pliny the Elder, noting that

Studius invented the landscape fresco genre during the reign of Augustus, lists the elements in such paintings as follows:

> villas, colonnades, ornamental gardens, sacred groves, woods, hills, fish-ponds, canals, rivers, beaches ... various views of people walking or sailing and, on land, visiting country houses on asses or in carriages ... people fishing, fowling, or hunting or even harvesting grapes ... famous villas with marshy approaches, tottering men carrying, on a bet, anxious women on their shoulders, and many further such lively and witty things.
>
> (*HN* 35.116)

This passage indicates that viticulture, at least, did appear in some of these paintings but there is no mention of olive or grain cultivation.[88] Vitruvius, writing in the latter half of the first century BCE, provides a similar description of the landscape genre, listing "harbors, headlands, seashores, rivers, springs, canals, shrines, sacred groves, mountains, cattle, shepherds" (*De arch.* 7.5.2, Morgan trans.) but nothing concerning arable farming.

Thus it is fairly clear that in the early Empire painters (or at least those who commissioned their work) had little interest in depicting some major aspects of rural life. On the one hand, it is not especially surprising that owners would want scenes of *leisure* decorating their villas. But, considering the substantial role of cereals in the Roman diet and the contemporary rhetoric concerning the virtues of farming, the almost complete invisibility of their cultivation is at least curious. Even with massive imports of grain from Sicily and Egypt and olive oil from Spain and North Africa, olive groves and fields of grain must have been common features of the real Italian landscape. Arable farming scenes do, moreover, appear in other media in the late Republic and early Empire. For example, a coin of Gaius Marius Capito, minted in 81 BCE, depicts, on the reverse, a farmer plowing with a yoke of oxen.[89] Plows, sickles, and other agricultural items appear on Republican coins as control marks.[90] Trajan's Column features a scene of Roman soldiers reaping.[91] A coin from Alexandria, minted during the reign of Antoninus Pius, depicts a harvester.[92] North African mosaics are a particularly rich source of rural agricultural scenes. The Zliten mosaics from late first century CE Libya feature scenes of hoeing, plowing, and threshing. The second century Neptune mosaic from Chebba in Tunisia shows a man harvesting wheat next to the personification of Summer who holds a sickle, and the El Djem Calendar Mosaic from late second or early third century CE Tunisia features scenes of rustic laborers. So Roman artists did produce more realistic depictions of country life, just not in this particular medium and region.

Although Cicero and other elite writers clearly considered manual labor to be vulgar (*Off.* 1.150), it is not the case that Roman artists and their employers had an aversion to scenes of labor. A fresco from the shop of

Verecundus at Pompeii depicts cloth-making while another fresco depicts cupids and psyches as *unguentarii* (makers of perfume). Cupids sometimes appear to make viticulture seem endearing but I am unaware of an example of cupids plowing fields or harvesting grain.

Scholars have studied the social context of Roman art extensively and the landscape genre has certainly not escaped their attention. Eleanor Leach has proposed that the rural poetry of Tibullus inspired some of these paintings.[93] Of the landscapes with Egyptianizing elements, Ling suggested that they were created "in response to a demand from educated patrons familiar with Alexandrian poetry."[94] More generally, Paul Zanker speculated that the Pompeian landscape frescos "provide[d] inviting locales to which the owner's thoughts could stray."[95] However, while Roman writers considered agriculture to be a teacher of virtue, wealthy Romans of the early Empire in general seem to have regarded *arable* farming as an especially ugly business and therefore its depiction as an unsuitable element of decoration. Pliny the Elder and Columella make this attitude clear even as they reiterate traditional praise for farming. Pliny writes that "fettered feet, criminal hands, and branded faces," slaves imprisoned in the *ergastulum*, do the farming in his day (*HN* 18.21). Columella implies much the same thing, saying that wealthy Romans of his day assign agricultural work to the worst of their slaves "as if delivering him to the executioner" (*Rust.* pr.3). In a slightly later passage he notes that "if a rich man has purchased a farm, from the crowd of his attendants and litter-bearers, he banishes the most worn-out in years and strength" (*Rust.* pr.12). Despite praising the farmer, even Cicero regarded digging and plowing as sordid work (*Fin.* 1.3).[96] If wealthy Romans tended to associate arable agriculture with worn out bodies and slave labor, we can hardly be surprised that they did not want to view depictions of it in their villas. Furthermore, even Romans who fully subscribed to the idea that agriculture instilled important virtues in its practitioners may well have felt that it was dangerous or simply depressing to announce the fact permanently on the walls of their houses. Viewers might interpret, for example, a fresco of Cincinnatus at his plow as an indirect attack on the emperor's monopoly on power. If that seems implausible, it is worth recalling the political climate in Rome during the reign of Tiberius. In 15 CE the quaestor Caepio Crispinus had charged his praetor, Granius Marcellus, with treason and testified, in support of the accusation, that Marcellus had placed a statue of himself higher than those of the Caesars (Tac., *Ann.* 1.74). At the trial, the emperor became furious when he learnt that Marcellus had cut the head of Augustus from a statue and replaced it with one of Tiberius. The Roman elite took notice: the funeral procession of Junia Tertia, half-sister of Brutus and wife of Cassius, in 22 CE, featured, as tradition dictated, the *imagines* of many noble ancestors, but those of the tyrannicides were conspicuously absent (Tac., *Ann.* 3.76).

The lack of depictions of arable agriculture in landscape frescos obviously need not have such an extreme cause as fear of treason charges. It could simply have been depressing for many Romans to contemplate that many of the

Republican civic virtues associated with farming were no longer necessary under the Principate. Scenes of viticulture evoked the pleasures of wine, images of livestock the charms of bucolic poetry or the relative ease of the shepherd's life, but plowing and harvesting called to mind only the ugliness of contemporary agriculture and the now useless virtues it promoted. Elite Romans of the early Empire had no wish to be reminded of the harsher aspects of farming. If, as seems likely, the cultivation of cereals and legumes was one of the least profitable parts of Roman agriculture, it is all the more understandable that the elite would not want depictions of it on the walls of their houses and villas.[97]

1.4 Conclusion

The many limitations of the textual and archaeological evidence are disheartening. Willem Jongman concluded that it was impossible to "reconstruct Italian agriculture from the remaining literary and archaeological data."[98] However, we can, I would argue, more closely approximate its economic dynamics by paying particular attention to what our sources, especially the agricultural writers, say about the specific activities and equipment required for farming and considering their implications. After discussing the constraints (climatic, demographic, botanical, etc.) on Roman farming in Chapter 2, I will turn, in Chapter 3, to the question of rural demand. I will show that nearly all farmers must have relied on external supplies to a certain degree and that smallholders would be especially dependent on them. I will also argue that we must consider the scale of rural demand for goods and services not directly connected to agricultural production, processing, and storage. Chapter 4 then examines the many ways farmers could acquire the money to purchase the various essential and non-essential goods they demanded. They could sell foodstuffs (from wine and grain to honey, cheese, and meat) in nearby towns (or further afield) but also sell their labor to larger estates with seasonal needs or even supplement the urban work force. There would also have been a local, rural market for some agricultural products. Of course, farmers could also acquire money through borrowing, an option that is also considered in Chapter 5, which examines the economic networks farmers inhabited. Farmers could turn to the market but they had other sources of support, including landlords, patrons, friends, and family. Cultivating relationships as well as crops gave them access to more assistance in times of great need. Having witnessed proscriptions, confiscations, and redistributions in the late Republic and further confiscations under the emperors, few Romans would be foolish enough to place too much faith in their property rights. Similarly, in an unpredictable Mediterranean environment, few possessed such large and dispersed landholdings that they could handle, solely out of their own resources, the droughts, damaging storms, and other disasters which invariably struck somewhere. Doubtless most Roman farmers sought economic security (along with other things), they just didn't seek it by means of an ideology of market avoidance. The sixth chapter, finally, connects the economic behavior of farmers to the broader economic and political history of Roman Italy in the late Republic and early Empire.

Notes

1 Garnsey and Saller (2015, 107) estimate that "80–90 per cent of the population worked in agriculture and accounted for 70–80 per cent of the value of production." Garnsey and Woolf (1989, 154–5) propose about 80 percent of the population "occupied the land." Temin (2001, 180) suggests 75 percent. Scheidel (2007a, 80) argues that "there is no good reason to believe that more than one person in eight would have been permanently or predominantly engaged in non-agrarian labor." See further discussion in Morley (1996, 182–3) and Erdkamp (2005, 12–13). Of course, some who did not directly work the land indulged in "the fantasy of playing the farmer" (Thibodeau 2011, 73).

2 For example, Flohr 2013a on fullers; McGinn 2004 on prostitutes; Bond 2016 on auctioneers, mortuary workers, tanners, and mint workers; Holleran 2012 on retailers and wholesalers.

3 For example, on peasants, see de Ligt 1990 and 1991; on *vilici*, Carlsen 1995; on aristocrats, Rosenstein 2008. On tenancy, see, for example, Kehoe 1997 and 2007b.

4 Ideology of self-sufficiency: Duncan-Jones 1974, 37 and Whittaker 1985, 58. See also, for example, White 1970, 394; Percival 1976, 161; Frayn 1979, 27; Dyson 1985, 77; Meijer 1990, 14–15; Paterson 1998, 165; Laurence 1999, 99; Horden and Purcell 2000, 151; Morley 2000, 216; Mattingly 2006, 288; Rosenstein 2008, 19; Harvey 2010, 709; Bodel 2011, 317; Holleran 2012, 241; and Witcher 2016, 475.

5 Peasant goal: Maróti 1976, 109; Knapp 1977, 13; de Ligt 1990, 43; Lirb 1993, 268; Morley 2007a, 92; Decker 2009, 229; and Marzano 2009, 40.

6 Hopkins 1980, 104. See also De Cecco 1985, 820; Thompson 1988, 214; Hopkins 1995/6, 61; Lo Cascio 2007, 627; and Bransbourg 2011, 98.

7 There are a number of passages from a range of texts that seem to reflect a general Roman disapproval for those who bought what they could have produced: Cicero, *Pis.* 67; Horace, *Ep.* 2.167ff.; Martial, *Ep.* 4.66.5 and 7.31; Varro, *Rust.* 1.59.2; Virg., *G.* 4.133; Ovid, *Ars Am.* 2.265–6; Plin., *HN* 19.57. Ovid advises lying about the market origins of some gifts of produce.

8 For example, Bowman and Wilson 2009, 48.

9 Autarky: Erdkamp 2005, 61 and 101.

10 For examples of local or regional conceptions of real or hypothetical self-sufficiency, see Campbell 1996, 91; Hopkins 2000, 263; Bang 2007, 29; and Morley 2007a, 19.

11 For example, Martin 1971, 41; Rawson 1976, 93; and de Ligt 1993, 110. This seems to be the sense in which Bannon (2009, 156) uses the term. *Autoconsumo,* a word employed by some Italian scholars in similar contexts (e.g., Marcone 2002, 327), is ideal for this type of practice.

12 For example, Wiedemann 1981, 97.

13 Specialized production: Witcher 2016, 468.

14 Horden and Purcell (2000, 272) argue that "to aim at subsistence is suicidal."

15 Economic rationality: Halstead 1987, 86.

16 Necessity: Carandini 1983, 195.

17 Rational response: Morley 2007a, 26.

18 For example, Knapp 1977, 13.

19 Frayn 1979, 150.

20 Weekly market day: Andreau 2002, 116–7.

21 Evans 1981, 441. Duncan-Jones (1974, 38) argued that the "self-sufficient estate ... owed part of its justification to the technological backwardness of the Roman economy." Laurence (1999, 99–100) persuasively critiques these ideas.
22 Frederiksen 1970, 336.
23 Veyne 1979, 269.
24 Myth: for example, Hopkins 1995/6, 61 and Paterson 1998, 158.
25 Mirage: Hopkins 2000, 257 and de Ligt 1993, 131.
26 Rhetoric: Horden and Purcell 2000, 115.
27 Ethical tenet: Horden and Purcell 2000, 272.
28 Holleran 2012, 242.
29 On the assumption of rationality, see Goodchild 2013, 200.
30 Behavioral economics: Thaler 2015.
31 This seems to be clearly reflected in the Roman agricultural texts. Varro's, of course, take the form of dialogues. In the others there is frequent quotation, discussion, and criticism of earlier agricultural writers.
32 On the role of friends and money in achieving security, see Hollander 2016.
33 See my previous discussion of rural demand for money: Hollander 2007, 122–135.
34 White (1973, 456) doubted we can know exactly when Cato wrote the work.
35 Dalby (1998, 227) considers part of the cabbage section to be a later addition.
36 White 1970, 19.
37 Spurr 1986a, 167.
38 Bodel 2012, 48–9.
39 Terrenato 2012, 87.
40 Terrenato 2012, 80–1.
41 Roselaar, forthcoming. On the question of *ergastula,* see Bodel 2011, 319.
42 Bodel 2012, 52.
43 Reay 2012, 61 and 67. See also Reay 2005.
44 Terrenato 2012, 85–6. Kron calls the *De agricultura* "a canny combination of homespun farm wisdom and sound agricultural practice, well calculated ... to enhance his well-honed public image among the many small farmers from whom he drew his political support" (2017, 119).
45 Bodel 2012, 53.
46 Henderson (2002, 131) calls it an "absurdist cast of agricultural names on etymo-legs."
47 White 1970, 24.
48 Kronenberg 2009, 74.
49 White 1973, 476. He calls the section on grafting "a glorious muddle." See also Green 2012, 33. Skydsgaard (1968, 181) notes many mistakes in Varro's work but concedes that he "does present us with a useful view of farming."
50 Brunt 1972, 154.
51 White 1970, 27.
52 Columella also wrote one extant book on trees, the *De arboribus.*
53 Von Stackelberg 2013, 1679. Henderson (2002, 113) calls Columella a "systemic organizer to reckon with, concerned throughout to articulate, waymark, and hypertext his individual lessons into an effective programme of teaching."
54 Martin 1971, 384.
55 Marcone 1997, 27.
56 Healy 1991, xxiv.
57 Frank Goodyear quoted in Wallace-Hadrill 1990, 80.
58 Thurmond 2006, 22.
59 Wallace-Hadrill 1990, 85.
60 The work was apparently considered valuable in Pliny the Elder's own time as, according to Pliny the Younger (*Ep.* 3.5), a certain Licinius once offered to buy the *notes* for 400,000 sesterces.
61 Dyson 2012, 130.

62 Kron 2015, 165. He also argues (164) that "the spread of large livestock" suggests that "the works of the leading agronomists are strongly representative of, and influential in shaping, the practices of farmers." Witcher (2016, 462) proposes that "the most reasonable approach is to accept that the writers had broad familiarity with agricultural practice, and that we can recognize details of specific agricultural tasks, such as vine pruning, but that knowledge of these processes does not imply that they were practiced in pursuit of the wider objectives promoted by 'characters' such as Scrofa."
63 See Spurr (1986a) for a defense of Virgil's agricultural knowledge. However, Horsfall (2001, 41) mocks the idea that the *Georgics* were "written to teach farming."
64 Doody 2007, 180 and 184.
65 Thibodeau 2001, 175.
66 See Uden (2010) for a discussion of this issue in poems about Priapus. The article also addresses the problem of the idealized self-sufficiency of the rural poor.
67 Kron 2017, 117.
68 Tools from Boscoreale: Harvey 2010, 697.
69 Jashemski 1979, 243–50.
70 Settefinestre: De Sena 2013, 6179.
71 Economy of Settefinestre: Carandini 1980. However, Marzano (2009, 33–4) raises doubts about the identification of the slave quarters.
72 Marzano 2007, 130.
73 Roman Peasant Project: Ghisleni et al. 2011, 101; Vaccaro et al. 2013, 129; and Bowes et al. 2017.
74 Amphorae: see, for example, Paterson 1982 and Peacock and Williams 1986.
75 For Pompeii, Jashemski and Meyer (2002) provide a comprehensive treatment including analysis of paleosols, pollen samples, timber, and plants, as well as fauna. Costantini and Giorgi (2009, 125), however, note that "archaeobotanical evidence for the agricultural economy of Rome and the surrounding area from the archaic through to the imperial period is limited." On the limitations of pollen data, see Bowes et al. 2017, 187–8.
76 Land pertaining to a farm: Frederiksen 1970, 331. See also Witcher 2016, 462.
77 Bowes et al. 2017, 177.
78 Witcher 2006, 49–50.
79 Launaro 2011, 83.
80 Small farmsteads: Rathbone 2008, 323. Garnsey and Woolf (1989, 154–5) believe peasant dwellings "have largely disappeared." Against this proposition, see Kron 2017, 125 n. 53.
81 Centuriation: Dyson 2003, 60.
82 Regional differences: Alcock and Cherry 2004; Launaro 2011, 77ff.
83 Field survey and the Gracchi: Marzano 2007, 130.
84 For example, side-by-side depictions of a maritime villa and food (Museo Archeologico Nazionale di Napoli Inv. 9391), and the garden fresco in the House of the Marine Venus.
85 For example, the famous fresco from the Casa del Centenario (IX.8.6) depicting Vesuvius features vineyards on the slopes of the mountain. Jashemski et al. (2002, 171) note that "the grape is one of the plants most often represented" in Pompeian wall paintings and list two depictions of vineyards from the city.
86 Fresco from the House of Triptolemus: Jashemski et al. 2002, 166–7.
87 Ling 1991, 143–4.
88 There is an apparent depiction of a *trapetum* or olive mill in a fragment of a fresco from the Villa Arianna at Stabiae (Inv. 64835).

89 Crawford #378. Crawford (1974, 392) notes that he does "not believe that there is any reference to Sulla's colonies or to his enlargement of the *pomerium*." For discussion of plowing scenes from Roman Anatolia, see Takaoğlu, forthcoming.

90 Examples of plows as control marks can be seen in the issues of L. Titurius Sabinus (Crawford #344) and C. Allius Bala (Crawford #336); the control marks of M. Plaetorius Cestianus (Crawford #405) include a sickle, crook, and ladder.

91 Agriculture in Roman art: Shaw (2013) discusses several examples such as the spectacular Tomba della Mietitura from Isola Sacra (see Baldassarre 1990), which features mosaics depicting plowing, hoeing and weeding, reaping, the transport of grain to the threshing floor, threshing, and winnowing. On representations of wild game, see Kron 2017, 129.

92 Coin from Alexandria: D.2989.

93 Leach 1980, 69.

94 Ling 1991, 143.

95 Zanker 1998, 189.

96 Similarly, Sallust (*Cat.* 4) refers to tilling the land as a servile task.

97 Kron (2015, 161) suggests that the decision to supply Rome (and other major cities) with cheap grain encouraged farmers near such cities to switch to "intensive mixed farming and the production of cash crops."

98 Jongman 2003a, 112.

2   The parameters of Roman agriculture

2.1 Introduction

In order to understand the economic behavior of Roman farmers, it is essential to first have an understanding of the physical environment in which they operated. The purpose of this chapter is survey the parameters of Roman agriculture including climate, geography, demography, and especially botany. For Italy, much like the rest of the Mediterranean,[1] the overall picture is one of considerable variety with a wide range of soils, climatic conditions, crops, access to irrigation, means of transportation, and so forth. All the agricultural writers emphasize this variability in some fashion. In his advice to Romans shopping for a new farm, Cato recommends taking note of the climate (*caelum*), weather, soil, geographical situation, labor supply, water supply, and access to transportation (*Agr.* 1.2–3). Varro has Scrofa describe regional differences in climate, contrasting the hills with the lowlands (*Rust.* 1.6), while Columella's extensive discussion of soils is worthy of a pedologist (*Rust.* 2.2). The implication is clear: Roman farmers were not to make the mistake of considering Italian agricultural conditions homogeneous.

2.2 Climate and geography

According to the climate classification system initially developed by Vladimir Koppen, most of Italy is in a Cs (i.e., temperate) zone with warm, dry summers and most precipitation occurring during the winter.[2] The climate of late Republican and early imperial Italy was roughly similar to that of twentieth-century Italy,[3] with, as now, many pockets of variation. Robert Sallares notes that "it is quite possible to have subtropical vegetation on the south side and temperate vegetation on the north side of the same mountain."[4] Nevertheless, with some changes in technique or timing, a farmer could grow many of the same crops and raise the same sorts of animals in most of the peninsula.[5] There was, however, also plenty of interannual variation. For example, during most winters Rome experiences no snowfall and temperatures tend to stay comfortably above freezing, but in January of 1985 the city endured six inches of snow and "temperatures fell to 20 degrees, the lowest recorded in a century."[6] Extreme weather can have a devastating impact on agriculture. Tacitus notes that storms did considerable damage to Campanian agriculture in 66 CE, destroying villas, orchards, and crops (*Ann.* 16.13). The eruption of Etna in 44 BCE seems to have caused significant cooling during the following two years and may have led to famine in Egypt.[7] As Horden and Purcell point out, both drought and too much precipitation could lead to problems.[8] Adequate drainage was often essential. Even a moderate amount of rain might cause trouble if it came at an inconvenient juncture, limiting the harvest time or soaking already harvested grain before it could be safely stored.[9]

Roman farmers were acutely aware of both regional and interannual variations and the agricultural writers tend to avoid being too rigid in their instructions about when to do critical tasks.[10] They also encourage estate owners to observe the techniques of their more successful neighbors (or the previous owner) if they are new to an area. Varro, for example, in a discussion of how to determine the number of slaves an estate needs, provides the recommendation that one should consider how the other farms in the neighborhood are staffed. The speaker declares that "with respect to agriculture, nature granted us two methods: experimentation and imitation" (*Rust.* 1.18.7).[11]

As Helen Goodchild notes, geological and climatic diversity make "simple zonation" difficult in peninsular Italy.[12] In general terms, however, there were differences between north and south, east and west, and the coast and higher elevations. The south is somewhat warmer and drier than the north; the coasts tend to be warmer than the interior; and the region to the west of the Apennine mountains, finally, in addition to being greater in overall area in comparison to the generally narrow eastern coastal plain, tends to receive more moisture. But within any given region there can be considerable variation. Indeed, soil scientists believe that "pedodiversity is probably higher in the Mediterranean region than in any other climatic zone."[13] The presence or absence of mountains, hills, rivers, streams, swamps, and forests all have an impact on the soil while the underlying stone helps determine the nature of groundwater. Human activities, of course, have a further impact (e.g., deforestation or drainage projects). While irrigation was not necessary for cereal cultivation, it was important for many other aspects of agriculture. Hence access to springs and the construction of aqueducts could have a considerable effect on the productivity of the land.

Despite the similarity between the climates of Roman and modern Italy, there were differences. The study of the Roman climate is very much in its infancy but there seems to be an emerging consensus that there was a period of relative warmth in the late Republic and early Empire. This "Roman Climate Optimum" or "Roman Warm Period" began in the third century BCE and lasted until the second century CE.[14] The warming peaked in the first century CE when it also started to become drier. As Saskia Hin has pointed out, this probably promoted Roman agriculture, accelerating plant growth, allowing plants to grow at higher elevations and further north, and making "marginal land less marginal."[15] While there were still outlier years, a recent synthesis of several types of evidence concluded that "[e]xceptional climate stability characterizes the centuries of the Roman Empire's rise,"[16] coming to an end in the latter half of the second century CE.[17]

There may also be significant differences between the modern Italian landscape and the one the Romans faced. In addition to the well-known issue of sea level,[18] Roman Italy may have featured more ponds and lakes as well as differences in the nature of rivers and floodplains.[19] Then as now, the Mediterranean climate promoted erosion and thus change in the conditions of agriculture.[20]

2.3 Demography

Demography remains a highly contested area of Roman history with a great deal of uncertainty surrounding fundamental questions. For example, *millions* of people separate the 'low count' estimate of the population of Roman Italy at the end of the Republic from the 'high count.'[21] Nevertheless, some basic points seem reasonably clear. To begin with, the urban population of Italy grew substantially in the late Republic. Jongman suggests that it tripled.[22] The city of Rome grew from around two hundred thousand to perhaps as many as one million inhabitants in the last two centuries BCE.[23] While urban conditions may not have been as bad as they are sometimes made out to be,[24] urban mortality was relatively high compared to the countryside and cities had to draw on the rural population to maintain their populations.[25] Hin notes that warmer periods tend to cause population growth but that the Roman Warm Period would also have improved conditions for malaria.[26] Despite the city's dangers, there were many appealing aspects to urban life which would have tempted country dwellers: subsidized or free grain, distributions of money and food, spectacles, and work opportunities to name just the most obvious "pull factors."[27]

It should be noted that there was likely considerable movement between city and country. The urban elite, of course, spent a good deal of time outside the city (and complaining about urban life) but there would be seasonal movements of poorer city dwellers as well. Many would go into the countryside to work as itinerant harvesters. Brent Shaw discusses the huge labor demands of the cereal harvest in Roman North Africa, for example, estimating that "something on the order of tens of thousands of men must have been on the move each summer to do the reaping."[28] Even if grain cultivation was less important in Italy, larger estates would still usually need to supplement their rural labor supply during harvest periods for any crop they grew in significant quantity. Witcher, furthermore, notes that "in pre-industrial societies townsfolk tend to remain deeply embedded in rural life, and individuals and families may move back and forth seasonally or through the lifecycle."[29]

Life expectancy at birth was between twenty and thirty years[30] and, when combined with the tendency of men to postpone marriage until their twenties or later and then marry girls in their teens, the majority of surviving children were fatherless by their mid-twenties.[31] This meant a potentially narrow window in which fathers could pass on their knowledge of farming to a son or sons.[32] While there is not much evidence for extended families living together, Richard Saller speculates that this practice may have been more common in rural areas.[33] This might have mitigated the effects of paternal mortality upon human capital. However, it is important to acknowledge that many of our demographic inferences are based on urban data and rural conditions may have been considerably different.[34] Partible inheritance would have tended to divide up family farms over the generations while several factors, especially the workings of debt law and upper-class demand for real estate, would, at the same time, have concentrated land in the hands of wealthier Romans. There was clearly a tremendous increase in the number of slaves in Italy during the late Republic, allowing wealthy landowners to exploit extensive holdings. Scheidel suggests an "influx" of from two to four million in the last two centuries BCE.[35] Less certain is the extent to which natural reproduction helped maintain the slave population, relative to the continued importation of new slaves from abroad.[36] The 'rescuing' of exposed infants certainly contributed to some extent.[37] How were slaves distributed between city and countryside? Here again it is hard to tell. On the one hand, there is ample evidence for the large scale use of slaves in agriculture, but it is also clear that the wealthy maintained sizeable urban slave staffs.[38] The high level of uncertainty about these issues makes it difficult to get a good sense of the nature of the rural labor supply.

2.4 Roman crops

Roman farmers cultivated a wide variety of crops. These crops often required different tools for cultivation and processing as well as different amounts of labor at different times, different amounts of water, and they would have different effects upon the soil in which they grew. Therefore, some familiarity with the major crops of Roman agriculture is essential if we want to understand the economic behavior of Roman farmers.

2.4.1 Cereals

The most important cereals cultivated by the Romans during the period under consideration were emmer, durum, and bread wheat. Unfortunately, compared to the other major crops, the cultivation, processing, and transportation of cereals did not produce much archaeological evidence.[39] Storage requirements, however, were substantial and remains of granaries do survive.[40] Emmer (*Triticum dicoccum*), referred to as *far, ador, adoreum,* and *far adoreum* in Roman sources,[41] was an early staple. It has a hulled grain and so, while requiring additional processing (e.g., parching) beyond the threshing floor prior to consumption, emmer was better protected from diseases and pests than naked wheats.[42] As Columella notes (*Rust.* 2.9.3), it could grow in a variety of soils,[43] and did better than barley in the damp conditions of central Italy.[44] Pliny the Elder says that emmer was the most commonly grown wheat (*HN* 18.81), praising its ability to withstand extremes of heat and cold (*HN* 18.83). It was sown in the fall, harvested in the summer, could have "a surprisingly high yield,"[45] and is high in protein.[46] Sallares calls it "the hulled wheat best suited to Mediterranean climatic conditions."[47] Less ideal for making bread than *triticum aestivum*, emmer had declined somewhat in popularity by the late Republic and early Empire but nevertheless remained an important crop.[48] The Romans used emmer to make *puls* (a kind of porridge or soup) and some kinds of bread, as well as employing it in important religious rites (Plin., *HN* 18.10).

Durum or hard wheat (*Triticum durum*) is a free-threshing or naked cereal, meaning that it does not require additional processing after threshing and winnowing before the grain can be milled into flour.[49] However, it lacks some of the protection afforded by a hull and so is more vulnerable to pests. Durum handles drought and hot weather well but is not suited to regions with cold winters.[50] While used to make bread, it is low in gluten, difficult to mill into fine flour, and so now tends to be used for pasta rather than bread.[51] There is some ambiguity about whether the use of the word *triticum* in Roman texts always refers to durum wheat as opposed to bread wheat,[52] and in many instances their remains are indistinguishable.[53] Nevertheless, it is likely to have been a significant crop in several parts of Roman Italy and probably was the cereal imported from Egypt to Rome in large quantities under the Empire.[54]

Bread or soft wheat (*Triticum aestivum*) was the other naked cereal grown by the Romans, who referred to it as *siligo*. It shares with durum the vulnerabilities associated with free-threshing wheat but was more resistant to cold weather and less tolerant of drought; thus it would have been cultivated more than durum in northern Italy.[55] Bread wheat, as its name suggests, produced the highest quality bread (Columella, *Rust.* 2.6.2; Plin., *HN* 18.86). Pliny the Elder notes that *siligo* had to be harvested quickly, a task made more difficult by its uneven ripening (*HN* 18.91).

Barley (*Hordeum vulgare*), though less popular than emmer, durum, and bread wheat, was nevertheless widely grown and of considerable importance. As a crop it has many advantages. Compared to wheat, it requires less water and can handle alkaline and saline soils better.[56] It is nutritious, disease resistant, tolerant of more extreme temperatures,[57] less wearing on the soil,[58] cheaper to produce in terms of labor input,[59] and grows quickly. There are two varieties: two-row (*distichum*) and the more widespread six-row (*hexastichum* or *cantherinum*) barley. Like most cereals, six row barley would be sown in the fall but one could sow the two-rowed variety between January and March (Columella, *Rust.* 2.9.16), a very useful feature should a farmer have been unable to sow a cereal crop in the fall or if that crop was already showing signs of failure. The barley harvest began in June (Columella, *Rust.* 11.2.50). However, barley has its drawbacks, particularly in places like central Italy whose damper soils were less conducive to its growth.[60] Furthermore, because it is usually hulled, after threshing barley needs to be roasted or pounded before it can be consumed.[61] Barley was not a preferred cereal for human consumption in the late Republic. There are numerous references to barley rations being used as punishment for Roman soldiers (e.g., Polyb. 6.38 and Suet., *Aug.* 24.2) who normally enjoyed emmer. Several passages from Cato's *De agricultura* indicate that barley was a standard crop (e.g., *Agr.* 35, 144, and 136) but, while it appears as an ingredient in some remedies and as food for hens and geese (*Agr.* 157 and 89), even Cato's slaves eat *triticum* (*Agr.* 56). Varro lists barley alongside emmer and *triticum* in a passage recommending the amount of seed to be sown per *iugerum* (*Rust.* 1.44), but his frequent remarks about it are almost always in the context of animal feed (e.g., 2.7.11 and 3.10.5–6). Columella says barley was better for animals than *triticum* and, for humans, quite helpful in times of famine (*Rust.* 2.9.14). He recommends a mixture of *triticum* and two-row barley as food for the *familia* (*Rust.* 2.9.14).[62] Pliny the Elder indicates that, at least by the first century CE, barley was typically animal feed (*HN* 18.74), though he does praise the medicinal qualities of barley flour (*HN* 18.74–5, 78).

The Romans also grew a variety of other cereals, although these were apparently of less importance. There seems to be wide agreement, for example, that einkorn (*Triticum monococcum*) played only a minor role in Roman cereal cultivation.[63] Pliny the Elder, who refers to it as *tiphe,* mentions it just twice, suggesting that it was foreign to Italy (*HN* 18.81 and 93). The other agricultural writers omit all discussion of einkorn but its remains do show up, albeit in relatively small quantities, in archaeobotanical studies of Roman-era sites in Italy.[64] Einkorn has some appealing qualities, including the ability to grow in poor or damp soils, and, being a hulled wheat, it stores well.[65] Although used to make a porridge and as fodder, einkorn yields were poor and it produced an inferior bread.[66] Another relatively minor cereal was rye (*Secale cereale*). Pliny notes its high yields as well as its ability to grow on all kinds of land and enrich the soil (*HN* 18.141).[67] In the same passage, however, he also complains of its bitterness and says that, even when mixed with emmer, it was disagreeable to the stomach. Neither Cato, Varro, nor Columella mention *secale* at all but it was certainly grown in northern Italy and there is some evidence for its cultivation in the south.[68] Similarly, spelt (*Triticum aestivum ssp. spelta*) is a hulled wheat that, though well suited to the land around Rome and good for making bread, was apparently not extensively cultivated by the Romans.[69] Although they are nutritious and can thrive in Mediterranean conditions,[70] the Romans may not have consumed oats (*Avena sativa*) to any great extent either. Pliny notes that the *Germans* liked to make a kind of porridge out of them (*HN* 18.149) but they do not feature prominently in the agricultural texts. Cato refers to oats as weeds (*Agr.* 37.5) while Columella mentions their use as fodder (*Rust.* 2.10.25 and 32) and says that they were sown in the fall. Pliny, however, says they are sown in the spring at least in northern Italy (*HN* 18.205). Both Columella and Pliny cite Virgil's claim that they damaged the soil (*Rust.* 2.13.3, *HN* 17.56, *Georg.* 1.77). Spurr concluded that they "were not part of the Roman diet" but Rosenstein suspects them to have been part of the repertoire of Italian smallholders.[71] Rice (*Oryza sativa*), though not common in Italy by any means, was apparently cultivated to some extent along the Po river in northern Italy.[72]

Millets, finally, deserve special mention.[73] In a 1983 article, Spurr described their many virtues: drought resistant, easy to reap, storable for relatively long periods of time, high in carbohydrates, "excellent animal forage," and possessing medical and veterinary uses.[74] Their drawbacks, however, were several: low yields,[75] sensitivity to the cold, the need for additional processing (parching), poor bread-making qualities (i.e., lack of gluten), and a detrimental effect on the soil requiring subsequent manuring or fallow.[76] But millets grow quickly and thus could be sown in the spring or even summer.[77] As an "emergency crop,"[78] millet was even more useful than barley, and had its own "agroecological niche."[79] Cato mentions *milium* along with *panicum* (a closely associated cereal) twice in lists of crops to be sown alongside such plants as turnip and garlic (*Agr.* 6 and 132). Varro mentions *milium* four times but *panicum* only once; he believed millet could keep for a hundred years if stored properly (*Rust.* 1.57). Both *milium* and *panicum* appear frequently in Columella's *De re rustica*, most notably together in a list he provides of especially pleasing and useful plants, which also includes peas, lentils, hemp, flax, and barley (*Rust.* 2.7.1). Pliny the Elder brings up millets quite often. He notes that some people preserve fruit in *milium* (*HN* 15.63) and that many people recommend burying a toad in the middle of a field of millet in order to protect the crop from sparrows and worms (*HN* 18.158).

What are the economic implications of Roman cereal cultivation? While we obviously do not have rainfall data from our period, it seems fairly certain that much of the Italian peninsula usually received enough precipitation to make irrigation unnecessary for cereal crops.[80] Different grains do, however, perform differently under different prevailing weather conditions and can require different amounts of processing with different equipment. A farmer who decided to concentrate on emmer, for example, because of the protections offered by its hulled grain, would have to expend resources on the additional processing. On the other hand, as Frits Heinrich points out, those who grew naked cereals might have to deal with "serious seed loss at the end of the ripening period."[81] He notes, however, that naked wheat chaff was more readily available for use as a commodity and hard wheat, being less bulky, was "easier to transport."[82] Different cereals thus would have more or less appeal to farmers depending on several factors including suitability of local growing conditions, intended use (domestic consumption or long-distance transport), level of demand (both for the cereal and the chaff), as well as available storage facilities and transportation options. Farmers who chose to grow a variety of cereals, either to minimize risk or to take advantage of different soils and situations, could need several different kinds of equipment and, ideally, separate storage areas for the crops. What about the role of differential cereal yields? There has been considerable debate over the fruitfulness of Roman wheat since it is an essential factor in calculations of carrying capacity and could help us determine the extent of official generosity in various land distribution initiatives. However, yields vary not only from species to species but also according to such factors as weather and soil quality as well as the amount of labor applied to cultivation. Rosenstein suggests that with "greater attention to hoeing and weeding," smallholders "could have increased yields dramatically."[83] Many farmers undoubtedly grew more than one kind of cereal (indeed, polyculture was probably typical) since this would allow them to take advantage of different soils, spread out their labor demands, and hedge their agricultural bets. However, this approach would tend to *increase* their need to spend time, resources, and potentially money on tools, processing, and storage.

2.4.2 Viticulture

As the source of the primary Roman beverage, viticulture was extremely important from a culinary, dietary, and economic perspective. The essentials of Roman viticulture can be briefly summarized. There were numerous varieties of *Vitis vinifera*, for wine-making but also table grapes and raisins, and the prudent farmer chose varieties suitable to local climate, soil, and situation (Columella, *Rust.* 3.1). Since all types of weather could harm at least some kinds of vines, Columella, our best source for Roman viticulture, recommends growing four or five varieties (*Rust.* 3.20). Presumably because they would not travel well, he advises against growing table grapes unless one was close to a city where they could be sold (*Rust.* 3.2). Viticulture required a significant investment (*Rust.* 3.3.8) and was quite labor intensive. It typically takes three years from planting to grape production.[84] The farmer needed ties to attach vines to trees, props, or trellises (Pliny *HN* 14.3, 17.199) and it was necessary to prune the vines regularly to promote proper grape production. To encourage the roots to penetrate deep into the soil, one had to trench around them every year (Cato, *Agr.* 33; Columella, *Rust.* 3.13). Irrigation might also be necessary. The large number of above-ground cisterns that have been documented from the environs of Rome may, in part, have been built to support viticulture.[85] Propagation, by means of cuttings (*viviradices*) or grafting, took both time and specialized knowledge. One also had to guard against a host of pests, including caterpillars (Cato, *Agr.* 95), mice, and foxes (Varro, *Rust.* 1.8.5). If you grew different varieties, they might ripen at different times, extending the fall harvest and pressing period (Columella, *Rust.* 3.21.5) but the simultaneous ripening of different varieties caused trouble as well since one would want to keep the different kinds of grapes separate during processing to avoid spoiling the wine (Columella, *Rust.* 3.21.6). While small-scale wine production did not necessarily require a large, expensive press and a dedicated press-room (*torcularium*), one certainly needed storage containers and at least some kind of vat for treading grapes.[86] Our sources, envisioning production on a fairly large scale, *did* expect such equipment and additional labor as well. Once the must (unfermented grape juice) was extracted from the grapes and stored in *dolia*, there were still things to do and problems with which to contend. It might be necessary to fix the taste of a wine (Cato, *Agr.* 109–10) or deal with pests (Columella, *Rust.* 12.31). Even before the harvest it was necessary to prepare equipment (*vasa, dolia,* baskets) and Cato felt the need to remind his readers to put away all the equipment afterwards, a good sign that such tools were numerous and, at least in the aggregate, expensive (*Agr.* 26). Given the capital and labor requirements of viticulture, it is not surprising, as Varro reports, that some believed it was not a good investment, with expenses devouring profits (*Rust.* 1.8). Columella has to argue at length in favor of the potential profitability of viticulture. He estimates the cost of a slave *vinitor* (vine-dresser), seven *iugera* of vineyards, osiers (willow shoots used to make baskets) and props, as well as two years' interest on their total price and calculates that even a lousy vineyard would have a greater annual profit than 6 percent interest on that sum (*Rust.* 3.3). He also emphasizes how the labor of the owner can pay off. He repeats the story of Paridius Veterensis who gave away two thirds of his vineyard as dowries for his two daughters but found that the remaining third was able to produce as much wine as the entire vineyard had previously (*Rust.* 4.3.6).[87]

2.4.3 Olives

Olive cultivation has received a great deal of attention in recent decades due to olive oil's prominent role in the Roman diet and the ample evidence we have for its processing (i.e., remains of press installations) and trade (e.g., Monte Testaccio, the hill at Rome made up of Spanish olive oil containers). While there were several varieties of olive tree (*Olea europaea*) and significant regional variation,[88] in general, they all had similar advantages as a crop. First of all, its main product, olive oil, was valuable as a nutritious food and fuel for lamps as well as for cleaning, to mention its most important roles.[89] David Mattingly estimates annual per capita Roman consumption at twenty to twenty-five liters.[90] (The Romans also consumed table olives.) Secondly, olive cultivation required relatively little labor and expense (except during the harvest and pressing in late fall and early winter) compared to other crops.[91] Lin Foxhall notes that, because the olives on the tree ripen gradually and are slow to rot, "picking can be spread out over several months,"[92] a considerable benefit to a farmer in need of flexibility in a farm calendar loaded with demands on labor. Annual pruning and digging (around the base of the tree) were, however, recommended (Cato, *Agr.* 32, 61). Thirdly, the by-products of olive cultivation were quite

useful. One could feed both pruned leaves and pressed olive residue to livestock.[93] *Amurca*, a "toxic olive juice,"[94] could be used to protect plants from
insects, kill weeds, fertilize the soil, preserve figs, protect clothes from moths,
rinse jars, and many other things.[95] Fourthly and finally, the olive tree can
handle brackish water, poor soil, and drought.[96] Once they reach maturity,
the trees can remain productive for centuries even if there are periods of
neglect.[97] On the other hand, olive trees take several years to reach full production,[98] and the olives require processing prior to human consumption.[99]
Propagation was not a simple matter since a tree grown from seed would
likely revert to a wild form.[100] Instead, it was necessary to graft shoots or
use cuttings. Furthermore, yield was very uneven from year to year, which
Mattingly has dubbed the "quirkiness of the oil harvest."[101] To compensate
for this, farmers often intercropped the olive orchard, growing things like
cereals between the trees.[102] With respect to olive oil's shelf life, opinions
vary. Mattingly suggests the oil would become rancid in two or three years
while Foxhall claims it could last more than six years in storage.[103]

There seems to have been a dramatic increase in olive cultivation (not
exclusively in Italy as Spain and North Africa also became major producers)
beginning in the late Republic and continuing into the early Empire.[104] The
importance of the olive is certainly reflected in the amount of space the
Roman agricultural writers devote to the crop. Cato, in addition to his scattered discussion of its cultivation, includes sample contracts for harvesting
and pressing of olives as well as one for the sale of the olives on the tree
(*Agr.* 144–6). Varro emphasizes the value of *amurca* which he claims was
poorly known (*Rust.* 1.55.7). Columella, of course, calls the olive "first of all
trees" (*Rust.* 5.8.1).

2.4.4 Other fruit trees

The Romans cultivated a variety of other fruit trees including fig, apple,
pear, plum, quince, pomegranate, carob, citron, and apricot. The cherry
and peach were introduced in the first century BCE.[105] Many of these trees
required grafting (e.g., apple, pear, plum, and cherry) to produce appealing
fruit. This would, of course, entail time as well as specialized knowledge
and equipment. There would often be a considerable wait before the trees
began to bear fruit. In the case of figs, for example, it would be three to four
years.[106] Nevertheless, fruit trees could be well worth the investment even
for poorer farmers. Columella reports that *rustici* (country folk or peasants)
often relied on dried apples (as well as figs) during the winter (*Rust.* 12.14).

2.4.5 Legumes

The Romans also grew many kinds of legumes (also known as pulses), including lentils, peas, chickpeas, beans, vetch, bitter vetch, fenugreek, and
lupins. These crops served several purposes. Some provided fodder for animals while humans consumed others. Legumes tend to be high in protein

and their roots add nitrogen to the soil, improving its fertility for subsequent crops.[107] The agricultural writers were well aware of their power in this regard (Cato, *Agr.* 138; Columella, *Rust.* 2.13.1–2). Columella lists lentils, beans, peas, chickpeas, and lupins among the legumes especially "pleasing and useful to man" (*Rust.* 2.7.1). While carbonized remains of legumes have been found at Pompeii and elsewhere,[108] in many instances they are very hard to detect.[109]

Lentils seem to have been particularly important, judging by the frequency with which they are mentioned. Cato provides instructions on their cultivation and preservation (*Agr.* 35, 116), as does Columella, who notes that they needed to be threshed, winnowed, dried, and stored, ideally in ceramic containers sealed with gypsum (*Rust.* 2.10.16). Pliny reports that lentils could be sown in the fall or spring, and grew well in a thin soil and dry climate (*HN* 18.123). He also notes their use in a number of remedies (*HN* 22.143). Varro recommends lentils for those keeping bees (*Rust.* 3.16.13).

Beans were another major crop. Pliny indicates that they were grown throughout Italy (*HN* 18.120), calling them the most important legume due to their potential use in bread-making but noting as well their role as fodder (*HN* 18.117). They had medicinal uses too (*HN* 22.140–1). Columella recommended growing *fabae* in the late fall but mentions an inferior variety (*fabae trimestres*) that could be sown in February (*Rust.* 2.10.8–9). After harvesting in the summer, beans were threshed and then dried (*Rust.* 2.10.12). While Cato and Varro imply that beans were a standard crop (e.g., *Rust.* 1.45), they have little to say about their cultivation.

Bitter vetch (*ervum*) was mainly used as animal feed since "its seeds are bitter and toxic to humans," but there are ways to remove the poisons so it could be eaten if absolutely necessary.[110] Cato notes that bitter vetch flour could be used to improve a harsh wine (*Agr.* 109). Columella and Pliny both mention its medicinal uses (*Rust.* 6.10.2; *HN* 22.151). Similarly, vetch (*vicia*) seems to have been mainly a fodder crop, appearing alongside fenugreek, bitter vetch, lupins, beans, acorns, and hay in Cato's lists of cattle feed (*Agr.* 27, 60). Pliny praises *vicia* for improving the soil, not requiring much labor on the part of the farmer, and having flexible sowing times, but he warns that it will damage vines if grown in a vineyard (*HN* 18.139). Lupins, by contrast, improve vineyards while also not demanding much effort to grow (*HN* 18.135). Columella says they "endure the farmer's carelessness" (*Rust.* 2.10.2). As for fenugreek, Columella calls it one of the best kinds of fodder along with *vicia* (*Rust.* 2.72). Pliny mentions its use as a deodorant (*HN* 24.187).

2.4.6 Fiber crops

Flax provided the Romans with several valuable products. Though it required intensive processing (drying, retting, further drying, pounding, combing, and spinning), flax could be turned into linen. The Romans had linen clothing, of course, but also sails and books. The fabric even had some agricultural applications. Linen cloth was used for straining (e.g., *HN* 18.77), bandages for humans and livestock (Columella, *Rust.* 6.11), and to make little seed packets (Columella, *Rust.* 11.3.32–3). According to Pliny, Cumaean linen made good fishing, fowling, and hunting nets (*HN* 19.10). Cato and Varro both mention the cultivation of flax, implying that it was a standard crop (*Agr.* 151.2, *Rust.* 1.22–3). Columella includes flax in a list of useful crops alongside, for example, lentils, beans, and barley, but warns against growing it unless yield and prices tend to be good in the region (*Rust.* 2.10.17). Sown in the spring and harvested in summer, flax grew quickly but damaged the soil (*HN* 19.7, cf. 18.205).[111] Pliny says that good flax was produced in the Po valley (*HN* 19.9), and remarks that spinning flax was *decorum* even for men (*HN* 19.18). The Romans do not seem to have used flaxseed oil.

Hemp also produced useful fibers, especially for agriculture. Columella instructs the farmer to sow hemp in late February or early March in "fertile, manured, and well-watered soil" (*Rust.* 2.10.21). Hemp had medicinal uses (*HN* 20.259), its seeds could serve as fodder,[112] and tools made from hemp, particularly ropes, seem to have been in high demand on the farm. Varro mentions mats (*tegetes*) made from hemp (*Rust.* 1.22.2) while Columella, in a discussion of the training of oxen, recommends the use of hemp ropes (*Rust.* 6.2.4). Varro refers to the use of hemp nets to keep birds in an aviary (*Rust.* 3.5.11). Rush (*scirpus*) had similar applications, being used to make coverings, mats, and candles (Plin., *HN* 16.178). Pliny also reports that peasants (*agrestes*) used the bark of several trees to make baskets and in the construction of their huts (*HN* 16.35).

2.4.7 Nuts

The Romans cultivated a variety of nut trees including almond (*Amygdalus communis*), walnut (*Juglans regia*), hazelnut or filbert (*Corylus avellana*), chestnut (*Castanea sativa*), and pistachio (*Pistacia vera*). Cato recommends growing walnut, hazelnut, and almond if one's property was close to a city (*Agr.* 8). The almond (*amygdala* or *nux Graeca* in Latin) could, of course, be sold as food but nut trees had broader commercial potential. Pliny says that almond oil served as a kind of soap, could, if mixed with honey, serve as a treatment for acne, and was also an ingredient in a number of other medicines (*HN* 23.85, 26.22). Walnut wood was an important building material (*HN* 16.223); walnut shells were used to dye wool while the nuts (*nuces iuglandes*) produced a red hair dye (*HN* 15.87). Pliny, always interested in the pharmacological, notes the medicinal uses of walnuts, including as an ingredient in one of Mithridates' famous antidotes (*HN* 23.147–9). He mentions that torches were made from the hazel tree (*HN* 16.75), and states that while hazel nuts (*nuces abellanae*) caused headaches and flatulence, they did have some medicinal uses (*HN* 23.150). For a farmer, chestnuts (*castaneae*) were useful as fodder, and Columella praises the tree, introduced into Italy in the first century BCE, for growing quickly from nuts and supplying props for vines (*Rust.* 4.30, 33).[113] Pliny notes the medicinal value of chestnuts too (*HN* 23.150), reporting that several varieties had been developed (*HN* 15.93), and that ground-up chestnuts were used to make a kind of bread (*HN* 15.92). According to Pliny, Lucius Vitellius introduced the drought-resistant pistachio to Italy (*HN* 15.91) from Syria (*HN* 13.51) late in the reign of Tiberius. He claims the nuts are useful in treating snakebites (*HN* 13.51, 23.150).

In terms of their labor inputs, some nut trees, such as pistachio and chestnut, require grafting for propagation but others can also be grown from seed (almond and hazelnut).[114] Walnuts can be grown from seed but grafting provides more consistent results.[115] Varro warns that walnut trees damaged the soil around them (*Rust.* 1.16.6) and notes that the nuts do not keep well (*Rust.* 1.67). Pliny advises against putting vines near hazel (*HN* 17.240). In an excellent illustration of the variability of growing conditions in Italy, he also observed that, while chestnuts and almonds did poorly around Rome, there were forests of them at the coastal city of Tarracina (*HN* 16.138). If the agricultural writers are any indication, the cultivation of nut trees increased somewhat in importance in the early Empire.[116] Cato mentions nut trees only twice and Varro's treatment is brief. He discusses almonds three times, twice in connection with apiculture (*Rust.* 3.23, 25), and recommends feeding acorns, walnuts, and chestnuts to dormice (*Rust.* 3.15), but makes no mention of hazelnuts. Columella does not mention pistachios but provides instructions for the cultivation of almonds, hazelnuts, walnuts, and chestnuts (*De arb.* 22, *Rust.* 5.10.12–14).

2.4.8 Vegetables and tubers

Since my focus here is neither on horticulture nor Roman foodways *per se*, I will keep my discussion of vegetables and tubers brief. Though not as significant as cereals, wine, and olive oil in the Roman diet, garden produce was nevertheless of considerable importance across the economic spectrum. Pliny the Elder makes this abundantly clear in Book 19 of the *Natural History*. On the one hand, we learn of the emperor Tiberius' obsession with cucumbers (*HN* 19.49), Nero's chive-based vocal remedy (*HN* 19.108), and the high prices choice vegetables could fetch (*HN* 19.54). On the other hand, Pliny calls the garden (*hortus*) the poor person's *ager* or farm (*HN* 19.51). Simulus, the pitiful farmer of the pseudo-Vergilian *Moretum,* sells most of the vegetables from his *hortus* in town while using some of its more ordinary produce (e.g., garlic and parsley) to add flavor to his meals.[117]

The Romans grew a vast array of vegetables and tubers in their gardens. These included garlic, leeks, onions, lettuce, cabbage, turnips, beets, celery, parsnip, melons, carrots, and asparagus. With an adequate water supply, it would have been possible to grow them year-round at Rome and elsewhere in Italy (at least in the lower elevations),[118] although the agricultural writers do give ideal planting times for many of these crops. In his discussion of lettuces (*lactucae*), for example, Pliny notes that they could be grown throughout the year in irrigated and manured soil but were best sown between the winter solstice and the spring equinox (*HN* 19.130).[119] Only quite recently, during the reign of Augustus, had it been discovered that one could preserve lettuces in *oxymeli* (honey vinegar) (*HN* 19.128). Gourds not only provided food but could also serve as storage containers for wine or water (Plin., *HN* 19.71).

Vegetable gardens seem to have been particularly important on the outskirts of Rome where vegetables could be grown for the urban market (see Chapter 4). Of course, then as now people took pride in consuming and serving to guests the products of their own gardens.[120]

2.4.9 Other plants

Beyond the crops mentioned here, there were other plants (and fungi) that formed part of the Roman diet, some cultivated, some gathered from uncultivated areas. Marginal lands, unsuitable for arable farming, thus could prove extremely valuable to nearby smallholders (who also sometimes used them for pasturing livestock).[121] These foods included fruits, nuts, herbs, and mushrooms.[122] Farmers also cultivated some plants that were purely or mostly decorative, such as many kinds of flowers and other crops that produced ingredients for medicines, or seeds or leaves which added flavor to dishes but whose nutritive contribution was relatively small. For example, the Romans grew three kinds of poppy, white, black, and common. The seeds of the white poppy were used in some recipes (Cato, *Agr.* 79 and 84) while the black poppy produced opium. The common poppy also had medicinal applications (Plin., *HN* 19.168–9, 20.198–204) and poppies could play a role in apiculture (Varro, *Rust.* 3.16.13). Similarly, sesame, a crop that seems to have grown in prominence by the early Empire, played small but significant culinary, medical, and cosmetic roles. Sesame does not appear in Cato's work and Varro mentions it only once (*Rust.* 1.45.1). By contrast, it comes up fairly frequently in Columella and Pliny the Elder. Columella includes it in his list of "pleasing and useful" crops and says it was sown in late June in the wetter parts of Italy (*Rust.* 11.2.56). He gives its seed requirements per *iugerum* along with those of more common crops (*Rust.* 11.2.75) and mentions sesame seeds in a couple of recipes (*Rust.* 12.15.3 and 12.59). Pliny notes that sesame was a summer crop (*HN* 18.96), as well as discussing the medicinal value of sesame oil (*HN* 23.95), and its role as a perfume ingredient (*HN* 13.12). Since plants such as these did not form part of the core Roman crop package, I will not discuss them *seriatim* as I have, for example, the main cereal crops. Those that seem to have served as significant cash crops for some Roman farmers will, however, receive attention in Chapter 4.

2.5 Livestock

It is increasingly clear that livestock played a more substantial part in Roman agriculture than previously appreciated,[123] and that stockbreeding was often highly integrated with arable farming.[124] The main Roman farm animals were cattle, sheep, goats, pigs, and chickens. Such animals provided meat, wool, hides, milk, eggs, and manure. Donkeys, mules, and horses were also kept as work animals along with dogs to help manage herds, keep guard, and, no doubt, provide companionship too. According to Columella, every farm needed a donkey (*Rust.* 7.1.3). While climate did not prevent any of these animals from being kept in any particular locale in the way precipitation rates and temperature limit where certain plants can grow, there were areas where it was easier or more difficult to raise different kinds of livestock. For example, Columella says that pigs could be raised anywhere but notes that marshy land was better than arid, and woods (*nemora*) containing a variety of trees (e.g., oak, wild olive, hazel, and juniper) were the most convenient (*Rust.* 7.9.2).[125] As with crops, the care of particular animals required certain knowledge and equipment, and would place extra demands on a farm's supply of labor at certain times of the year. Diseases and pests could damage or kill animals so medical knowledge and supplies were essential. Some ingredients (e.g., old wine) in a given medicine would be readily available on most farms, but cinnamon, myrrh, frankincense, and sea-tortoise blood, the other ingredients in a potion Columella recommends for sick oxen (*Rust.* 6.5.3), were probably much more difficult to acquire (though we may reasonably doubt the widespread use of this particular remedy). Animals also needed to be housed and, judging from the attention devoted to their construction and maintenance (e.g., Varro, *Rust.* 2.4.14 on the pigsty), these facilities required considerable effort too.

The economic consequences of keeping animals should be clear: a ramifying network of material needs and labor obligations. Let us consider, for the purposes of illustration, raising sheep. A Roman farmer who actually attempted to be self-sufficient would need a small herd of sheep to supply wool for his family's clothing. He would need to spend time and energy monitoring and caring for the herd. He would need access to pastures – 'marginal' land may often have been available nearby – or have to dedicate some of his arable land to fodder crops and be able to store fodder as well.[126] (Transhumance, another potential strategy, would have its own set of drawbacks for the independent smallholder.) Columella notes that sheep are especially sensitive to temperature so the farmer would need a well-built pen too (*Rust.* 7.3.8). Shearing typically took place in spring or summer as did breeding, while lambing occurred in fall or winter.[127] Shears are not essential for shearing sheep but plucking the wool, which Varro says some still did in his day (*Rust.* 2.11.9), would be more time-consuming. The farmer would also need time and equipment to convert the wool into yarn and then woven fabric. Sanna Lipkin has estimated that it would take "400–520 hours to produce one mantle of wool."[128] After shearing, the farmer would also have to devote time to caring for the sheep's skin (see, e.g., Varro, *Rust.* 2.11.7). Indeed, Columella emphasizes that sheep required considerable knowledge of veterinary medicine (*Rust.* 7.3.16) and that they are susceptible to a number of diseases (*Rust.* 7.5). Of course, accomplishing all these tasks would not be an insurmountable task for larger and wealthier households but these demands would be more difficult or impossible for smaller *familiae* seeking to produce all their requirements for themselves. As will be discussed in subsequent chapters, they would have compelling reasons to turn to the market for iron tools to make shearing more efficient, for wool to obviate the need to devote resources to caring for a herd, or simply for clothing to conserve domestic labor for other tasks. Another option, perhaps quite common, would be to form partnerships with other farmers.

Roman farmers raised a wide variety of animals beyond those mentioned so far, everything from lampreys to dormice. As an increasing body of research now shows, the Romans undertook pisciculture on a very large scale. By the early Empire there were dozens of large "maritime fish farms" on the Adriatic and Tyrrhenian coasts.[129] It is also important to keep in mind that, just as wild plants supplemented the Roman diet, so too did wild animals. Studies of faunal remains are beginning to indicate that the level of consumption was surprisingly high.[130] Though such animals, especially fish, could form an important part of one's diet, they can hardly be considered staples. I will consider them further in Chapter 4 in a discussion of sources of agricultural income.

2.6 Conclusion

The factors discussed in this chapter are not, of course, the only constraints Roman farmers faced. Religious practices, social conventions, political imperatives, and market forces all helped determine what farmers did and did not do with their land and its produce. These are issues I will explore in subsequent chapters.

The Romans cultivated a wide variety of plants with the potential to provide them with a healthy diet as well as fodder, clothing, some tools, fuel, and medicine. Likewise, a variety of livestock produced a range of useful goods and, in some cases, labor power. This might seem to make self-sufficiency attainable but suitable land and water for these plants and animals were not evenly distributed across the landscape, even given the diversity of microclimates. Only the wealthiest could hope to own all necessary types of land. Everyone else's options would be much more limited. Furthermore, smaller farm households would be unlikely to have the labor and extensive knowledge to cultivate a wide range of crops and raise many different animals in an efficient manner. Even a smallholder with land suitable for subsistence needs of grain, wine, and olive oil, would also require trees for building material, making tools, and fuel, as well as the means to process and preserve the produce. Clothing was another necessity. If the smallholder had a small herd of sheep, there would be the need for additional pastureland or fodder, additional time spent on supervision of the animals, as well as the tools and time needed to convert wool into clothing. As I will make clear in the next chapter, our sources do not indicate that farmers were so self-reliant. Quite the opposite! They expect farmers to make lots of purchases and only sought to *limit* them.

Notes

1 As Garnsey (1988, 9) notes, "the climate of the Mediterranean is (and has always been) exceptionally diverse from region to region." At a sufficient level of "magnification" Horden (2013, 1582) describes the Mediterranean as "a world of virtually innumerable micro-climates, in which one side of a hill may be arid, the other side jungle." Horden and Purcell discuss Mediterranean "microregions" and "microecologies" at length in *The Corrupting Sea* (2000).
2 Climate classification: Torrent 2005, 418; Rombai 2002, xxxviii.
3 Italian climate: Thomas and Wilson 1994, 172; but not as warm as now: Hin 2013, 77.
4 Local variations: Sallares 2007, 15.
5 Goodchild 2013, 208.
6 Dionne, E. J. Jr. 1985. "Six Inches of Snow, Followed by a Roman Holiday." *New York Times,* January 8. Accessed January 21, 2017. www.nytimes.com/ 1985/01/08/world/six-inches-of-snow-followed-by-a-roman-holiday.html
7 Etna: Forsyth 1988. See also Hin 2013, 78–9.
8 Excessive precipitation: Horden and Purcell 2000, 180.
9 Bad timing: Shaw 2013, 24–30.
10 See, for example, Varro, *Rust.* 1.6.3; Columella, *Rust.* 11.1.30–2.
11 Similar advice is found at *Rust.* 1.19.2.
12 Zonation: Goodchild 2013, 199.
13 Pedodiversity: Torrent 2005, 424.
14 Roman Warm Period: Hin 2013, 74. See also Manning 2013, 163.
15 Hin 2013, 86–7. Chapter 3 of her *The Demography of Roman Italy* discusses Roman climate change and its implications at length. Roman agricultural conditions appear considerably less dire than once thought (see, for example, Evans 1981).
16 Climate stability: McCormick et al. 2012, 174.
17 McCormick et al. 2012, 203.
18 Sea levels: Pirazzoli 1976.
19 Floodplains: Thomas and Wilson 1994, 143; Brown and Ellis 1995, 69.
20 Erosion: Brown and Ellis 1995, 68.
21 Low count and high count: Launaro 2011, 21.
22 Urban population: Jongman 2003a, 106.
23 Population of Rome: Morley 1996, 39.
24 Urban conditions: Morley 2005, 198.
25 Urban mortality: Jongman 2003a, 106.
26 Consequences of the Roman Warm Period: Hin 2013, 89 and 92.
27 Urban 'pull factors': Jongman 2003a, 106.
28 Seasonal harvest labor: Shaw 2013, 23.
29 Embedded in rural life: Witcher 2017, 47.
30 Life expectancy: Scheidel 2007a, 39.
31 Fatherless children: Saller 2007, 92.
32 Passing on knowledge: Saller 2012, 74–5. On the dangers of ignorance in agriculture, see, for example, Columella, *Rust.* 11.1.28.
33 Extended families: Saller 2007, 92.
34 Urban demographic data: Scheidel 2007b, 400–1.
35 Slave population: Scheidel 2005, 64. See also the discussion in Kay (2014, 178–82) who assumes "steady growth" and estimates a slave population of 1.4 million in 28 BCE.
36 Sources of slaves: Scheidel 1997.
37 Exposure: Harris 1999.
38 Scheidel (2011, 289) believes "slaves were disproportionally concentrated in cities." He estimates six hundred thousand "non-farming slaves in late

Republican and early Imperial Italy" out of a total population of between 1 and 1.5 million. Although Bodel (2011, 312–13) emphasizes that "work was not the most important function a slave performed" and concedes (319) that the evidence for slave quarters on Roman rural estates is not especially strong, he argues (315) that "more slaves worked in agriculture than in any other activity."

39 Evidence for cereal cultivation: Witcher 2016, 467.
40 Granaries: see, for example, Rossiter 1978, 57–62.
41 Words for emmer: they are sometimes translated as "spelt." For discussion, see Spurr 1986b, 13 and Moritz 1955, 129–34. Moritz points out that, since *t. spelta* may not have been cultivated in the Mediterranean until the first century CE, it is best not to translate *far, ador,* and *adoreum* as "spelt."
42 Virtues of emmer: Braun 1995, 34–5; Decker 2009, 100.
43 On emmer and different types of soil, see also Carlà 2013b, 2390; Braun 1995, 34–5.
44 Emmer and damp conditions: Sallares 2007, 31; see also Columella, *Rust.* 2.6.4
45 Yield: Braun 1995, 34–5.
46 Protein content of emmer: Costantini and Giorgi 2009, 133.
47 Sallares 2007, 31.
48 Importance of emmer: Braun 1995, 34–5; White 1995, 39; Rosenstein 2004, 47; Zohary et al. 2012, 40; Carlà 2013b, 2390. But see Heinrich (2017, 156) for the virtues of flat breads.
49 See Heinrich (2017, 150) for discussion of the advantages of naked over hulled cereals with respect to processing.
50 Durum and climate: Spurr 1986b, 15.
51 Durum bread: Hanelt and Institute of Plant Genetics and Crop Plant Research 2001, 2574; Sallares 2007, 32.
52 Interpretation of *triticum*: Spurr 1986b, 15–17; Decker 2009, 102.
53 Costantini and Giorgi (2009, 133) note that, in the absence of "their diagnostic rachis segments," one cannot distinguish durum from bread wheat so it is often only possible to be certain of the presence of a free-threshing wheat on a site.
54 Hard wheat: Zohary et al. 2012, 40–7; Heinrich 2017, 153.
55 Bread wheat: Spurr 1986b, 15; Sallares 2007, 32; Costantini and Giorgi 2009, 134; Heinrich 2017, 160–1.
56 Barley: Zohary et al. 2012, 52; Goodchild 2013, 203.
57 Virtues of barley: Decker 2009, 105.
58 Impact of barley on the soil: Carlà 2013a, 1050.
59 Labor requirements of barley: Sallares 2007, 31.
60 Barley and damp soils: Goodchild 2013, 203.
61 Processing barley: Decker 2009, 104–5.
62 For further discussion of barley as a food for slaves, see Braun 1995, 33–4.
63 Einkorn: Spurr 1986b, 13; Costantini and Giorgi 2009, 133; Sallares 2007, 31–2.
64 Archaeological evidence for einkorn: Costantini and Giorgi 2009, 148.
65 Virtues of einkorn: Zohary et al. 2012, 34; Costantini and Giorgi 2009, 133.
66 Einkorn yields and bread: Zohary et al. 2012, 34.
67 Zohary et al. (2012, 59) add that rye has "winter hardiness," is drought resistant, and "succeeds under conditions in which wheat frequently fails."
68 Rye: Spurr 1986b, 13; Costantini and Giorgi 2009, 148. Jasny (1942, 763–4) described rye as "unadapted to the Mediterranean region."
69 Spelt: Costantini and Giorgi 2009, 131–3. See also note 41 above.
70 Virtues of oats: Zohary et al. 2012, 66.
71 Cultivation of oats: Spurr 1986b, 14; Rosenstein 2004, 47.
72 Rice: Zohary et al. 2012, 74.
73 Broomcorn or common millet (*Panicum miliaceum*) and Foxtail or Italian millet (*Setaria italica*) are *milium* and *panicum* in Roman texts.

74 Virtues of millets: Spurr 1983, 1–14.
75 Millet yields: Sallares 2007, 33.
76 Disadvantages of millets: Spurr 1983, 4–14.
77 Growing millet: Spurr 1983, 7; Decker 2009, 109; Zohary et al. 2012, 69.
78 Emergency crop: Decker 2009, 107–8.
79 Niche crop: Sallares 2007, 33.
80 Irrigation: Goodchild 2013, 199–200; Witcher 2016, 470. Thomas and Wilson (1994, 189) suggest that around Rome grain, olives, and vines would not usually need more water than was provided by precipitation. See also Forni 2002, 112.
81 Seed loss: Heinrich 2017, 155.
82 Advantages of naked wheats: Heinrich 2017.
83 Yields: Rosenstein 2004, 74. See Heinrich (2017) for a discussion of crop-selection decision-making in Roman Italy. He concludes (169) that "Ironically, in terms of cereal crop-selection, the most profound effect of the globalization of the Roman economy for Italian rural communities was perhaps the absence of the pressure to change."
84 Time to grape production: Zohary et al. 2012, 121.
85 Cisterns: Wilson 2008, 749.
86 See White (1975, 164–5) on the *linter,* a "hollowed-out tree-trunk" used for treading grapes. This would be a low cost, if relatively inefficient, means of processing.
87 Pliny also seems sensitive to the issue of profitability in viticulture (e.g., *HN* 14.25) and sings the praises of Acilius Sthenelus who famously turned around a couple vineyards (*HN* 14.48–50).
88 Olive trees: see, for example, Columella, *Rust.* 5.8.3; Mattingly 1996, 218–19.
89 Olive oil: Mattingly 1996, 222–3; Jongman (2007b, 603) notes that olive oil "makes up for some deficiencies of a cereal-dominated diet."
90 Olive oil consumption: Mattingly 1996, 223.
91 Labor requirements: Columella, *Rust.* 5.8.1; Spurr 1986b, 134–5; Mattingly 1996, 221.
92 Olive harvest: Foxhall 2007, 126. Pliny (*HN* 15.13–14) notes that different varieties ripen at different times.
93 Byproducts of olive cultivation: Foxhall 2007, 82.
94 *Amurca:* Foxhall 2007, 138.
95 On the uses of *amurca,* see, for example, Cato, *Agr.* 91–103; Varro, *Rust.* 1.55.7; Columella, *Rust.* 11.2.30.
96 Resilience of olive trees: Mattingly 1996, 214–15; Foxhall 2007, 6.
97 Endurance of olive trees: Columella, *Rust.* 5.8.1–2; Mattingly 1996, 216–18; Zohary et al. 2012, 116.
98 Foxhall (2007, 76), writing about Greece, says it takes twenty-five to thirty years while Mitchell (2005, 93–4), discussing Asia Minor, says as many as fifteen years. Mattingly (1996, 219), however, notes that fruit production might *begin* as early as the fifth year. See also Pliny the Elder's discussion (*HN* 15.3) of variation in the time between planting and olive production.
99 For olive oil, processing involved pulping and then pressing the olives, followed by the separation of the oil from the lees and *amurca.* See Mattingly 1996, 229; Thurmond 2006, 73–110; and Foxhall 2007, 11.
100 Propagation: Zohary et al. 2012, 117.
101 Yield: Mattingly 1996, 225.
102 Intercropping: Goodchild 2013, 206.
103 Shelf life of olive oil: Mattingly 1996, 225; Foxhall 2007, 80.
104 Trends in olive cultivation: Mattingly 1996, 216–18; Hitchner 2002, 72.
105 Cherry and peach trees: Horden and Purcell 2000, 259.
106 Fig trees (time to production): Zohary et al. 2012, 126.
107 Virtues of legumes: Zohary et al. 2012, 75.

108 Remains of legumes: Jashemski and Meyer 2002; Costantini and Giorgi 2009.
109 Detecting legumes: Wilson 2009, 215.
110 Bitter vetch: Zohary et al. 2012, 92.
111 Gleba (2004, 33) says that in the south of Italy flax was planted in late fall.
112 Hemp: Zohary et al. 2012, 106.
113 Zohary et al. 2012, 150; Horden and Purcell 2000, 259.
114 Propagation of nut trees: Zohary et al. 2012, 147–52.
115 Walnuts: Zohary et al. 2012, 149.
116 However, Bradley (1987, 49) cautions that, if you judge by the omissions and inclusions of the agricultural writers, you can easily make erroneous conclusions.
117 Of course, given that Simulus has a slave as well as a team of oxen, and regularly comes home from the market heavy with coin, he cannot be especially poor. For discussion of this poem and the genre to which it belongs see Kenney 1984.
118 Year-round vegetable production: Thomas and Wilson 1994, 158–9.
119 Similarly, Pliny (*HN* 19.137) says that cabbage was sown throughout the year but the best time to sow it was at the fall equinox.
120 This is the flipside of the disapproval expressed towards those who purchased what they might have grown themselves. See Chapter 1, note 7.
121 On the concept of marginality, see Horden and Purcell (2000, 178–82) who note (182) that "the areas we dismiss as least hospitable ... are among the most diverse and complex portfolios of complimentary productive opportunities."
122 Gathering wild plants: Witcher 2016, 471.
123 Witcher (2016, 462) notes that "compared with the cultivation of crops, much less attention has been devoted to the role of animals in ancient farming."
124 Kron (2000) has made the convincing case for widespread intensive (ley) farming in Roman Italy, that is, agriculture in which the pastoral and arable elements are closely connected and mutually beneficial. The synergy of plants and animals could lead to tremendous yields, as Columella indicates in the preface to the sixth book of his De re rustica, where he turns from plants to animals. The Roman Peasant Project has recently discovered what they call "the first coherent body of data that may indicate the practice of convertible agriculture by rural small holders" in southern Tuscany (Bowes et al. 2017, 199). The excavation of other small rural sites in Italy may eventually make clear how common ley farming was, in what regions, and when farmers came to adopt its methods.
125 The same went for sheep. Columella (*Rust.* 7.2.3) discusses the kinds of sheep best suited to the plains, hills, and mountains of Italy. He also notes that Greek or Tarentine sheep require more food and attention than other breeds and so were hardly profitable to raise unless the master was present (*Rust.* 7.4.1).
126 Pasture lands: Roselaar (2010, 207) suggests that "Public pasture lands were probably available everywhere, either as *subseciva* or as unassigned land in the surroundings of the town."
127 Shearing time would also vary with the breed of sheep. See, for example, Varro, *Rust.* 2.11.6 and Columella, *Rust.* 7.4.7. On the mating season, see Columella, *Rust.* 7.3.11.
128 Wool working labor estimate: Lipkin 2012, 109.
129 Fish farms: Kron 2015, 166.
130 Consumption of wild game: Kron 2017, 128–36.

3 Buyers and borrowers

The rural demand for goods, services, and money

3.1 Introduction

All the ancient agricultural writers acknowledge the need for the farmer or estate manager to make use of the market, and not just as a place to sell surplus produce for a profit. Despite stating that the *paterfamilias* should be a seller, not a buyer, Cato makes this clear in the very same passage, recommending the owner review with the *vilicus* the accounts of cash, what is needed, and what services they should purchase (*Agr.* 2.5–6). He goes on to state that the *vilicus* should not make purchases without the knowledge of the owner (*Agr.* 5.4) but, again in the same passage, acknowledges that he will be hiring laborers. When, late in the *De agricultura*, Cato discusses the duties of the *vilicus,* he reiterates this point, listing purchases as one of those duties (*Agr.* 142). As for the *vilica*, Cato warns against her being too *luxuriosa*, implying that her responsibilities included at least some purchases. Varro paints a similar picture. His dialogues acknowledge that some farms would not have access to the necessary raw materials to make certain tools (*Rust.* 1.22.2). The character Stolo advises that such things be purchased for use rather than for show (*ob speciem*). This is an intriguing admonition, suggesting that some farmers took considerable pride in their tools, perhaps engaging in a form of rural conspicuous consumption.[1] Stolo further suggests that the tools purchased be good, cheap, and from nearby. As for Columella, while he does argue that there is no point in having vineyards if you have to buy props (*Rust.* 4.30.1), in his discussion of the *vilicus*, he too indicates that purchasing was part of the job, stating that the *vilicus* should not go to town or the *nundinae* except to sell things or purchase necessities (*Rust.* 11.1.23–4). Farm purchases were to be limited, not eliminated.[2] Markets were essential to the practice of agriculture. In a speech delivered in 63 BCE, Cicero claimed that one of the reasons the Romans had not destroyed Capua during the Second Punic War was precisely so that the farmers in the region would have a place to buy equipment (*Leg. Agr.* 2.88–9). At Capua and elsewhere, the *nundinae* were periodic markets occurring every eighth day. They seem mainly to have taken place in towns, although they might also be held on a private estate (Plin., *Ep.* 5.4, 13). They gave farmers a chance to sell goods and make purchases but were also seen as an opportunity for them to learn about public affairs (Columella, *Rust.* 1.pr.18; Macrob., *Sat.* 1.16.34). According to Seneca, it had once been the only day of the week on which country folk washed themselves thoroughly (*Ep.* 86.12). In addition to rural smallholders, merchants and bankers could also be in attendance. While in some areas the *nundinae* seem to have been scheduled on different days in different cities, presumably so that merchants could travel between them, they are not thought to have been the places where the large surpluses of major estates were sold.[3]

Given that even the larger farms discussed by the agricultural writers had to make some purchases, smaller farms may have had even greater needs. The purpose of this chapter is to evaluate rural demand for goods, services, and coinage. That means, first and foremost, what farmers needed in order to operate their farms that they could not make themselves or do themselves (at least without great difficulty). But we need to consider not just what we view as necessities. It is also important to consider what Roman farmers purchased that was not strictly-speaking necessary for agricultural production. There is no reason to think that farmers were immune to fashion, inured to hard labor, or embraced asceticism (though idealizing sources sometimes portray them as such). Obviously the needs of farmers would vary considerably based on the size, nature, and location of their landholdings, their access to public land, overall wealth, social status, and ambition as well as, of course, their choices in terms of agricultural strategy (that is, whether they embraced the market and produced cash crops or tried to produce more of what they consumed themselves). Despite these variables and the dearth of price data, we can get a sense of the likely shape and some features of rural demand. I have divided the goods and services necessary for cultivation into three categories: 'start-up' supplies (that is, the materials and hired expertise necessary to establish a farm, e.g., building supplies and agricultural implements), seasonal supplies (e.g., additional labor), and maintenance supplies like the services of a blacksmith or doctor.[4] The categories are not absolute, of course, since one can think of certain items that might easily fit under two different headings. For example, a farmer with a large vineyard or olive orchard might consider the possession of processing equipment like mills and presses to be essential while small-scale producers and likely tenant farmers might pay to make use of another's equipment and so avoid the capital outlay. Smallholders might also share equipment with neighbors to minimize such expenses.[5]

3.2 Start-up requirements

Start-up supplies constitute the largest group of potential expenditures since they could include not just the tools of cultivation, processing, and transportation but also slaves, livestock, buildings to house family, slave staff, and livestock, storage facilities, containers, and cisterns. But, of course, farmers would rarely start completely from scratch on untouched land. Except in some instances of colonization, the individual prospective farmer would already presumably have at least some of the necessary equipment (brought from a previous location or inherited) and the land would have some useful structures. Before we turn to the components of startup demand, it is worth contemplating Cato's shopping advice, inexplicably inserted between discussions of the harvest sacrifice of a pig and share cropping agreements:

> At Rome: tunics, togas, coats, patchwork cloaks, boots. At Cales and Minturnae: hoods, iron tools: knives, spades, mattocks, axes, harness, *murices,* chains. At Venafrum: spades. At Suessa and in Lucania: carts. At Trebla Alba and Rome: vats, tubs. Tiles from Venafrum. Roman plow will be good for vigorous soil, Campanian ones for grey earth. Roman yokes will be the best. A detachable plowshare will be the best. Olive crushing mills at Pompeii, Nola and the wall of Rufrium. Locks, keys and bolts at Rome. Buckets, oil urns, water pitchers, wine urns and other bronze vessels at Capua and Nola ... Hoisting ropes and all goods made of esparto at Capua. Roman straining bags at Suessa and Casinum.
>
> (*Agr.* 135, Dalby trans)

Cato seems to have a thorough knowledge of the market, mentioning almost a dozen different places to buy goods. He also gives us a good sense of the sheer number and variety of equipment a farm required: clothing, iron tools, harness for animals, vehicles, ceramic storage containers, processing equipment, baskets, and ropes. Cato's dependence on the market should not be overstated, of course. In an earlier passage he does suggest that the farm staff make baskets (*Agr.* 31), as do the later agricultural writers.

3.2.1 Building supplies

Perhaps the single biggest startup cost, though no doubt the most infrequent, would be the actual construction of the farm. Cato discusses villa construction early on in the *De agricultura*. A hired *faber* did the building and his fee was based on the number of roof tiles employed (*Agr.* 14).[6] It is worth noting that Cato envisions the *faber* as building not only sheds, stables, housing for slaves, the hearth, doors, walls, and windows but also things like benches, seats, looms, and presses. The owner provided the building materials, some of which, for instance the roof tiles, were to be purchased. Unmentioned but also essential would likely be a well, cistern, or some means of storing water for household use. In the neighborhood of Rome, Thomas and Wilson have suggested the "quality and uniformity of cistern construction ... perhaps argues for construction by professional groups rather than by unskilled navies under the direction of the estate owner."[7] Certainly the use of hydraulic cement would require ingredients and expertise that the average farmer was unlikely to possess.[8] As usual, we are much less well informed about the nature and cost of the houses of smallholders.[9] We should envision considerable variation in the scale and nature of construction. In some cases, farmhouses, like the houses of Pompeii, would have been built with a considerable amount of iron.[10]

3.2.2 Metal agricultural tools

Agriculture obviously does not require the use of iron tools since it was practiced for millennia prior to the advent of the Iron Age but numerous pieces of evidence indicate Roman farmers commonly employed a variety of iron implements. First and foremost, we may appeal to Virgil and Ovid who both use plowshares-into-swords imagery to contrast peace and war. Virgil writes of curved sickles melted down into a sword (*G.* 1.508) while Ovid has rakes transformed into a helmet and mattocks turned into javelins (*Fasti* 1.699–700).[11] The poets, along with our other agricultural writers, take it for granted that the farmer will have an arsenal of metal tools. Sometimes farmers could even be buried with their tools.[12] It is worth reviewing the main kinds of metal agricultural tools in order to give a sense of their variety and number. The tools a particular farm would use varied, of course, depending on soil, terrain, and crops grown as well as, it seems, regional traditions.[13]

The plow is the quintessential Roman agricultural implement, used to break up the soil prior to sowing but also to help preserve moisture in the soil and destroy the shallow, lateral roots of nearby trees.[14] The Roman plow seems to have been mostly wooden but the *vomer* (plowshare) was made of iron since it had to withstand the friction of being forced through the soil.[15] Some soils might not require the use of a plow and sometimes the terrain made their use difficult and farmers would have to resort to smaller, hand-held implements. The *ligo* (mattock), a two-handed iron tool, might take the place of a plow in hilly terrain and saw use in confined spaces such as gardens for weeding and working the soil around cultivated plants.[16] *Rastri* (drag-hoes) could also perform the functions of a plow as well as having several other important uses with respect to working the soil.[17] It is impossible to know what proportion of Roman farmers used plows. White noted the possibility that most peasant farmers used "manual implements only" or shared a plow though, he conceded, "of this we have no contemporary evidence."[18] It is perhaps telling, however, that when Virgil sets out to describe the farmer's tools, he mentions the plowshare first (*G.* 1.162). Regardless of whether it was absolutely essential, the plow, drawn by oxen (as well as sometimes donkeys and mules), allowed a farmer to cultivate more land in less time.

Sarcula (hoes) had several functions. Cato (*Agr.* 155.1) refers to their use in managing the flow of water, and Columella mentions them with respect to mixing manure into the soil (*Rust.* 2.15.2), covering seeds with earth (*Rust.* 2.10.33), and breaking up clumps of earth (*Rust.* 2.17.4). Pliny the

Elder says that "mountain peoples plow with hoes" (*HN* 18.178), as well as noting their use to work the soil in pastures (*HN* 18.186). White called the *sarculum* "among the commonest of all agricultural implements."[19] The *bidentes* was a two-pronged hoe used in vineyards, olive orchards, and to clear undergrowth.[20] *Capreoli* were apparently more specialized weeding hoes which Columella recommended for asparagus cultivation (*Rust.* 11.3.46).

'*Falces* constitute another large group of iron agricultural tools. These implements had curved blades and included scythes, sickles, some knives, and billhooks. White suggests twelve "apparently distinct types."[21] Among these are the *falx veruculata,* a kind of sickle used by some to harvest grain (Columella, *Rust.* 2.20.3); the *falx faenaria,* a scythe for hay (Cato, *Agr.* 10.3); *falces silvaticae* for use in arboriculture; and *falces vinitoriae* and *vineaticae* for dressing vines and harvesting grapes (Columella, *Rust.* 4.25; Varro, *Rust.* 1.22.5). There were also *falcula* and *falcicula.* White indicates that the former was used to harvest certain kinds of grain.[22] The *falx messoria* seems to have been a sickle used to harvest cereals.[23] Varro makes it clear that different regions employed different methods of harvesting grain and thus different tools (*Rust.* 1.50). Iron could even have a role in the processing of grain as Varro notes that the *tribulum,* a threshing sledge, sometimes had iron parts (*Rust.* 1.52.1).

Spades (*palae*) and shovels (*rutra*) were also essential farm equipment. Cato recommended four *palae* for the olive orchard (*Agr.* 10). Columella mentions the *bipalium,* "essentially a trenching implement" according to White, several times in connection with nurseries (e.g., 3.5.23), vines (11.2.17), and gardens (11.3.11). Cato recommended using the *bipalium* to manure fields while Pliny refers to their use in nurseries and vineyards (*HN* 17.159). Suggesting many poorer farmers used *bipalia*, Columella notes that the *rustici* called them *sestertia* (*De arb.* 15). As for shovels, Cato recommended having five *rutra* for a two hundred and forty *iugera* olive orchard and four for a one hundred *iugera* vineyard (*Agr.* 10.3 and 11.4). They were used to move and mix material such as manure and soil.

Farmers also needed hatchets (*dolabrae* and *dolabellae*) and axes (*secures*) for pruning in vineyards and orchards (Columella, *Rust.* 4.24.4–5; Cato, *Agr.* 10.3), clearing out roots while plowing fields containing trees (Columella, *Rust.* 2.2.28), cutting down trees, and dressing wood. These and other iron woodworking tools facilitated the making of wooden tools and baskets on a farm. Going without them would make other aspects of domestic production more arduous. For orchards, there were also pruning saws (*serrulae*) (Columella, *De arb.* 64). I have already mentioned the use of shears (*forfices*) with respect to sheep but they could also be used to prune vines or harvest grapes (e.g., Columella, *Rust.* 12.14.4; Plin., *HN* 15.62). That there were superstitions against the use of iron implements in the cultivation of crops such as basil and mint may suggest that such tools were fairly common even in the garden (see, e.g., Plin., *HN* 19.177).

Beyond the implements used directly in cultivation, Roman farms would also have a variety of other metal tools and utensils. Cato mentions chains (*catellae*), buckets (*hamae*), braziers (*foculi*), and assorted bronze containers (*Agr.* 11, 135); Pliny refers to iron combs used in the processing of flax (*HN* 19.17: *pectitur ferreis aculeis*).

Our sources, textual and archaeological, give the impression that iron agricultural tools were in widespread use among Roman farmers. While not essential, they made farming more efficient in terms of time and labor.[24] It is worth recalling that very few farmers would have the proper raw materials, fuel, expertise, and time at their disposal to make their own farm implements. Even if they possessed a furnace and experience working with metals, most farmers would have needed to purchase iron ore. Diodorus Siculus, writing in the first century BCE, says (5.13) that iron ore was mined on Elba, refined there by crushing and smelting into pieces looking like large sponges, and then sold to merchants who transported them to Puteoli and elsewhere where they were sold again and made into armor and agricultural tools which *other* merchants then took away to sell elsewhere. Elba, of course, was not Italy's only source of iron: iron ores (of varying quality) were found in many places, but Diodorus (and other sources) strongly implies that ironworking tended to be a specialized industry.[25] In his discussion of mill construction, Cato notes that the *faber* made the iron parts. There is little reason to think farmers regularly made their own tools though there are certainly references to them conducting maintenance. Virgil says that one of the tasks of the farmer during rain showers was to hammer back into shape the plowshare (*G.* 1.261–2), while Pliny the Elder lists the sharpening of *ferramenta* among predawn winter farm chores (*HN* 18.236). Columella says much the same thing and adds the making of new handles for such iron tools (*Rust.* 11.2.92). Columella also reminds his readers to acquire and sharpen *falculae* and *ungues ferrei* (iron hooks) prior to the grape harvest (*Rust.* 12.18.2). Cato does refer to a *fabrum ferrarium* but only as a place to hang grapes (*Agr.* 7.2). No doubt the space was also used to repair iron tools.

Virtually all farms and farmers would have purchased their iron tools. Our sources also make clear it that these tools constituted a significant investment. Iron implements were not luxuries but, in the aggregate, they were fairly valuable. We see this from the emphasis placed not just on their care and maintenance but also the need to guard them. Columella says the storehouse for tools should be close to where the *vilicus* and the *procurator* live and specifies having an enclosed space within it for the iron tools (*Rust.* 1.6.7). Columella advises the *vilicus* to frequently inspect the *ferramenta* (*Rust.* 11.1.20). He also warns that one of the dangers of a farm close to marshes was rusty equipment (*Rust.* 1.5.6). Varro has Scrofa recommend close supervision of the farm tools. The *vilicus* should keep them stored close by. There should be lists of them both in the city and on the estate. They should either be locked up or easy to keep an eye on (*Rust.* 1.22.6). Later legal sources, which discuss how to define the *instrumentum* of farms when they were bequeathed in wills, suggest farm equipment was quite valuable since they were clearly worth fighting over.[26] Unfortunately, it is difficult to move beyond a vague sense that iron tools were valuable to a more precise sense of their cost.[27]

3.2.3 Wooden tools

Of course, a great many farm implements were (or could be) made of wood. These included reaping combs (*pectines*), winnowing forks (*ventilabra*), fans (*valli*), and reaping boards (*mergae*), as well as some threshing sleds (*tribula*), spades (*palae ligneae*), rakes (*rastelli*), and pitchforks (*furcae*) which could also be of iron.[28] Due to their perishable nature, few examples survive.[29] Our sources do encourage the farmer to make or have made such tools on the farm but they were likely also for sale since not all farmers would have access to suitable timber or the time or skill to make them. As noted earlier, iron woodworking tools – which would have to be purchased – would greatly facilitate the making of wooden agricultural tools.

3.2.4 Livestock

Turning to livestock, established herds could, of course, be naturally self-perpetuating but the agricultural writers give ample advice on the purchase of such animals.[30] Veterans and urban dwellers settled on the land would have to purchase them (if they were not provided for them by the authorities organizing the settlement scheme). Breeding mules seems to have been a fairly specialized enterprise so most farmers would have had to turn to the market if they wanted to acquire one.[31] The same probably held for donkeys, a farm necessity according to Columella, who calls them inexpensive and common (*Rust.* 7.1.3). Given the widespread practice of pastoralism in conjunction with arable farming, it is likely that sheep, cattle, and goats could be acquired with relative ease within most agricultural communities. The same no doubt goes for dogs without which, Varro tells us, no villa was safe (*Rust.* 1.19.3).

3.2.5 Storage

The secure storage of valuable iron tools could require the acquisition of more iron tools in the form of locks and keys,[32] but farmers also had to protect their produce from a wide array of pests, especially insects. Ceramics played the major role in this regard. On the one hand, ceramic vessels were not absolutely necessary for olive oil and wine; skins and barrels could be used instead,[33] but archaeological finds indicate the widespread distribution and use of a vast array of pottery types, for storing wine and oil but also other produce, as well as for their processing and consumption. Leaving aside the many kinds of fine ware and lamps found in great quantities in most rural excavations, the major kinds of coarse ware include amphorae, *mortaria*, jars, lids, cups, bowls, dishes, jugs, and casseroles. Cato recommends having lots of *dolia* (*Agr.* 1.4) which could be used to store olive oil, wine, grain, lupins, fruit, as well as the useful byproducts of olive and grape processing.[34] Cato even suggests using them to salt hams (*Agr.* 162).

Some villas did have pottery workshops (*figlinae*) for the production of ceramics (including bricks and tiles).[35] Varro's Scrofa mentions a pair of agricultural writers who discussed their management but he concludes such workshops are not strictly speaking an aspect of agriculture (*Rust.* 1.2.22). Most farmers who used ceramic vessels, especially but not exclusively the poorer ones, would have had to purchase them (often from the manufacturers of local course ware).[36] Cato expected the farm would purchase jars (*dolia*), pots, and urns; he names Capua and Nola as the best places to buy wine and oil urns, and water pots (*Agr.* 135). With respect to grain storage, there was a wide range of options from purpose-built *granaria* to pits or, in the case of Simulus, a "poor heap of grain" (*Moretum* 16).[37] Poorer farmers thus seem to have had inexpensive storage solutions, although spoilage rates may well have been higher for the simpler methods.

3.2.6 Clothing

Turning to clothes, while Cato does suggest that slaves on an estate mend their own cloaks (*centones*) and hoods (*cuculiones*) (*Agr.* 2.3) and that old clothing be made into *centones* (*Agr.* 59), he notes that they can be purchased (at Rome, Cales and Minturnae respectively) along with coats (*saga*), wooden shoes (*sculponeae*), tunics, and togas (*Agr.* 135). He does this despite indicating, as already noted, the presence of sheep on the farm. Attesting to their value and importance, Columella wanted the *vilicus* to examine the slaves' clothing twice a month (*Rust.* 11.1.21). He seems to envision some domestic production of clothing (*Rust.* 12.3.6). In his discussion of the *vilica* he states that she should do wool work and so be able to demand it of others. He recommends that she make clothing for herself, the overseers, and the more reputable slaves in order to lessen the burden on the accounts of the *paterfamilias*. As discussed in Chapter 2, it would be difficult for smallholders to make their own clothes. Finally, it is worth noting Lucretius' statement that looms require iron (5.1351), indicating that domestic textile production itself seems to have necessitated purchases.

3.2.7 Processing equipment

Processing equipment, including mills, presses, ceramic vats, and lead cauldrons, was another area where a farmer might have to rely on the market. Depending on the needs of the individual farm, these could be permanent fixtures or merely seasonal requirements. In addition to the metal items already mentioned, Cato provides purchasing advice concerning ropes for presses, animals harness, and vehicles (*Agr.* 135). Varro, however, encourages domestic cultivation of hemp, flax, rushes, and esparto grass in order to make rope and other household necessities. One of the advantages of this domestic cultivation is the ability to sell supplies to neighboring farms (Varro, *Rust.* 1.16).

For those with substantial vineyards or olive orchards, mills and presses could be major expenses. Cato notes that one could buy a *trapetus* or olive-mill at Pompeii for 384 sesterces (not including transportation and assembly costs) (*Agr.* 22.3). Columella indicates that there were regional variations in olive processing equipment (*Rust.* 12.52.7). Smallholders did not necessarily need substantial equipment to process olives, grapes, or grain.[38] In a discussion of harvesting agreements, Cato refers to the possibility of hired harvesters making use of a communal mill, giving the miller the same fraction of the processed grain as the harvester received of the harvest (*Agr.* 136). Communal mills, of course, would allow agricultural workers or smallholders to obtain flour without a large mill of their own (they could also mill small amounts on an ad hoc basis with a quern).

3.2.8 Other equipment

Rounding out the startup requirements of a farm might be animal harness such as yokes for oxen or *plostra* (carts and wagons). Cato gives purchasing information for all these things (*Agr.* 135). Concerning slaves, certainly an integral factor in his conception of the villa, Cato does not provide purchasing advice but Varro and Columella have shared their opinions. Varro discusses how to select suitable farm slaves and notes the excellent reputation and high prices of slave families from Epirus (*Rust.* 1.17). Columella has more to say about the *treatment* of slaves than their purchase but does recommend investing in a good vinedresser (*Rust.* 3.3.8).

3.3 Seasonal requirements

With respect to seasonal supplies, again, much would depend on the size of the farm, its location, and crops. Smallholders and tenant farmers, as noted earlier, might make use of the processing equipment and facilities of others. Although here the evidence is rather limited, this might include mills, presses, and threshing floors. Payment would often have been in kind.

Harvest times required additional, short-term storage containers to move produce from the field or orchard to a long-term storage structure or processing facility. Baskets were ideal for this sort of task. Pliny (*HN* 16.35) notes the use of many different kinds of bark by the *agrestes* to make baskets for carrying grapes and other crops, and to thatch their huts (*tuguria*). Surprisingly, Cato lists places to buy *fiscinae*, small baskets made from the stems of broom, rushes, and twigs (*Agr.* 135.2–3). Varro, by contrast, recommends that the farmer plant thickets in order to have withes with which he could weave baskets and make other agricultural equipment (*Rust.* 1.23). Columella recommends making various kinds of baskets (if the area provided suitable raw materials) and Pliny says much the same thing (*Rust.* 11.2.10; *HN* 18.232–3). For the olive harvest, in addition to baskets, Columella mentions ladders, iron ladles, sponges, mats, and other things he says he is unable to remember (*Rust.* 12.52.8). He even suggests having extras to replace those lost during the harvest (*Rust.* 12.52.9).

Smallholders are thought to have had an overabundance of labor relative to the agricultural needs of their land,[39] but it is generally accepted that larger, slave-run estates had to hire seasonal labor especially during harvest times.[40] Additional workmen, rather than equipment, would be their main seasonal requirement. This is certainly the impression given by Cato who refers to hired laborers several times (e.g., *Agr.* 4). He advises against allowing the *vilicus* to hire the same worker two days in a row (*Agr.* 5.4), perhaps to limit his ability to incur major expenses.[41] Outside labor is envisioned for the grain and olive harvests as well as the processing of the olives into oil. With respect to the olive harvest there is even the suggestion that such workers might be able to capitalize on competition among estates for their labor (*Agr.* 144). While, as Launaro and others have argued, there probably was no "shortage of free labour in the Italian countryside between late Republic and early Empire,"[42] this situation may not yet have obtained in the first half of the second century BCE at least during harvest periods.

Smaller-scale wine producers who, as has already been noted, did not require a press,[43] may have made do with more primitive means of extracting grape juice or gained access to the presses of others by some means (perhaps exchanging the use of a press for a portion of the product).

3.4 Maintenance requirements

The third category of rural demand, maintenance supplies, encompasses an array of goods and services that would be needed on an even less frequent basis or irregularly. The agricultural writers make it clear that there were considerable upkeep expenses on a farm. Buildings decay; people and animals get sick; tools and containers break. Cato mentions lime along with wax, resin, sulphur, and gypsum as ingredients in a putty for mending cracks in *dolia* (*Agr.* 39). Lime may also have been used to treat acidic soil in some areas.[44] Farms with hives or lime-kilns or the appropriate kinds of trees could produce *some* but not *all* of these substances themselves. From time to time a farm would inevitably require the services of blacksmiths, artisans, and doctors. Varro, who sometimes appears to be a stronger advocate of domestic production than Cato, notes with approval that some farmers "like to have local doctors, fullers, and artisans on an annual retainer rather than on their own farms since sometimes the death of one such expert ruins the farm's profit" (*Rust.* 1.16).

Medical expenses, as will be discussed below, would be a major concern for many farmers but it is especially interesting that Varro mentions fullers in his list. This would seem to suggest that some estate owners kept sheep on such a scale that it made sense to accomplish all the processing of woolen textiles on site.[45] In the same passage Varro acknowledges that it sometimes makes better financial sense to buy certain supplies rather than making them oneself.

Although the agricultural writers all emphasize the importance of caring for tools, eventually new equipment became necessary. Columella says that a donkey is essential for every farm precisely because it can be used to fetch supplies from town (*Rust.* 7.1.3). He cites Virgil's lines about holidays in the *Georgics*: "often the farmer loads his sluggish donkey with oil and cheap fruits, returning from the city with a millstone or lump of black pitch" (*G.* 1.273–5).

Gum (*cummi*) was needed to coat olive oil *dolia* (Cato, *Agr.* 69 and Columella, *Rust.* 12.52.17) while both *dolia* and other containers were sometimes treated with pitch (*Agr.* 25). Although Pliny the Elder says that the best gum came from Egypt and cost three *denarii* per pound, inferior types were available more locally since a variety of plants produce it (*HN* 13.20). Pitch was used to coat vats and *dolia,* as well as to treat the hooves of oxen.[46] Pliny provides instructions on making pitch (from the pitch pine) (*HN* 16.52–3, 57–8) but says that in Italy the Bruttii made the best kind (*HN* 14.127, 16.38). Columella suggests twenty-five pounds of pitch were necessary to coat a *dolium,* recommending that a fifth part be of Bruttian pitch (*Rust.* 12.18.7). Virgil's description of a small farmer, quoted above, has him often return from town with a lump of black pitch. Other important substances that might not be available locally include bitumen and lead. Bitumen could be used to waterproof ceramics and baskets,[47] while lead was sometimes used to repair cracked *dolia.*[48]

Humans and most livestock require salt.[49] It was also an important preservative for meat, olives, and some stored fodder, as well as the key ingredient in brine.[50] Judging from Cato's instructions, a farm's demand for salt could be considerable: he recommends a *modius* (a peck or about two dry gallons) of salt for each slave per year (*Agr.* 58) and half a *modius* of salt for each ham to be preserved (*Agr.* 162). Pliny notes the importance of salt to livestock, milk production, and the quality of cheese (*HN* 31.88; see also Columella, *Rust.* 7.3.20). While not typically that expensive, few farmers would have been able to produce their own salt. With no sources of salt in the interior of Italy, many Romans would not have had easy access to this condiment and preservative.[51] This may explain why salt pans apparently played a significant role in early Roman history. Livy notes the establishment of salt pans at Ostia following its foundation (1.33) and mentions the high price of salt, and efforts to combat it, shortly after the expulsion of the Tarquins (2.9).[52] There were many different varieties of salt and Pliny the Elder mentions many sources (*HN* 31.73–7), but sea salt, the most readily available kind in Italy, required "extensive processing to become palatable."[53] Thurmond suggests that "much crude salt was sold ... with little or no refinement" and points out Cato's instructions for turning common salt (*sal popularis*) into white salt (*Agr.* 88).[54] Pliny (*HN* 31.83) notes that salt could be extracted from the *muria* of salted fish through boiling.

3.5 Extra-agricultural expenses

The rural inhabitants of Italy bought plenty of *non*-essential goods and services too (at least from a strictly agronomical perspective). On the high end, imported wine amphorae, glass, perfume and *garum* containers, and non-local marble regularly turn up in excavations of villas. Of course, not all these goods derive from purchases in the market. Some would have been gifts from patrons, clients, or friends, and some would have been produced on another property owned by the same person. Other kinds of acquisitions would leave little or no trace but we should imagine farms and villas in some cases having expensive clothing, scrolls, wooden furniture, and decorative items.

Religious expenses are likely to have been a substantial component of rural expenditure. In part this was due to the nature of agriculture. The risks and uncertainties involved in farming caused many to seek the help of the gods. Indeed, Virgil (*G.* 1.338ff.) emphasizes the need for the owner and his *pubes agrestis* (adult male farm staff) to worship the gods, especially Ceres, with sacrificial offerings and songs. There were a host of rituals associated with the agricultural calendar. Cato mentions several of these, starting with the obligation of the *paterfamilias* to salute the *Lar familiaris* upon arrival (*Agr.* 2) and including a *suovetaurilia* (the sacrifice of pig, ram, and bull) to purify land (*Agr.* 141). Columella, who, for example, specifies sacrifices to Liber and Libera prior to the vintage (*Rust.* 12.18.4), planned to devote an entire book to the subject of agricultural lustrations and sacrifices once he had finished his *De re rustica* (*Rust.* 2.21.5–6).[55] Pliny the Elder, who seems somewhat less interested in recording agricultural rituals, does pass along the recommendation that basil be sown with curses and insults to help it grow better (*HN* 19.120). While this could be accomplished without expense, it shows the extent to which some Romans viewed agriculture as religiously fraught.[56] In a discussion of storage facilities, Columella, echoing Xenophon's *Oeconomicus,* mentions the storage of tools and furniture but emphasizes setting aside the items used for religious activities, as well as women's and men's holiday clothing (*Rust.* 12.3.1). There were a number of rural festivals including the Paganalia, Compitalia, Robigalia, Fordicidia, Cerealia, Vinalia, Ambarvalia, and Sementivae.[57] Tesse Stek points out that coins themselves could play a role in some festivals, contributed by attendees to finance them but also used as a means of counting the participants.[58] Some Romans considered the *nundinae* to be festivals as well (Macrob., *Sat.* 1.16.28).

There was probably considerable rural demand for incense. Cato mentions it as an ingredient in two medicines and as part of the offerings to Ceres prior to the harvest (*Agr.* 70, 127, and 134). However, Pliny the Elder, alluding to the religious piety of Roman peasants in the introduction to his encyclopedia (*HN* pr.11), notes that it was not a problem if such people made offerings of milk and *mola salsa* rather than incense.

Cato and Columella are both concerned to limit the religious activities and expenditures of their estate managers. This seems to go beyond the issue of the observance of established, traditional festivals. Quite early in the *De agricultura* Cato says the *vilicus* should not perform rites except during the Compitalia or at the hearth (*Agr.* 5.3). He also warns that the *vilicus* not have any parasites and then specifies the *haruspex*, *augur*, *hariolus* (soothsayer), and *Chaldeus* (astrologer), all religious professionals who presumably expected to be paid in some fashion (*Agr.* 5.4). Columella expresses similar concerns, saying that the *vilicus* should not make sacrifices unless instructed by the master nor make the acquaintance of a *haruspex* or *saga* (fortune-teller) (*Rust.* 11.1.22). These recommendations come right before the advice that the *vilicus* avoid the city and the *nundinae* unless absolutely necessary. The *vilica* received similar instructions. Cato says she "should not perform religious rites nor order another to perform them for her unless commanded by the master or the mistress," and should "know that the master performs religious rites for the whole household" (*Agr.* 143). Among the important qualities of the *vilica*, according to Columella, is that she be *remotissima* from superstitions (*Rust.* 12.1.3). Of course, the unauthorized religious activities envisioned by the agricultural writers were probably not exclusively or primarily related to farming. Rural Romans, like their urban counterparts, had spiritual needs and they could be reasonably expensive. The Parentalia, a February festival in honor of ancestors, could involve elaborate feasts and offerings at their tombs, along with the burning of incense.[59] *Profusio* tubes for offerings at grave sites were fairly common. Ovid has to emphasize that small offerings were best (*Fast.* 2.533–42). Free smallholders would not have the threat of an owner to prevent them from indulging in costly sacrifices or the services of an astrologer. It is not unreasonable to suppose that religious devotions sometimes placed farmers in financial difficulty.

3.6 Borrowing and sharing

While farmers would have to resort to the market for many of the goods and services discussed in this chapter, they could, as has occasionally been pointed out, acquire some of them through loans, in-kind exchanges, partnerships, or simply the goodwill of their neighbors. These aspects of supply will be considered in more detail in Chapter 5 but it is worth noting here that these approaches would have been important strategies for many farmers. Cato emphasizes the value of one's neighbors, saying that, if they like you, you will be better able to sell your goods and hire workers; in reference to the issue of construction expenses it is notable that he says that friendly neighbors will help you build, providing labor, beasts of burden, and timber (*Agr.* 4). He also seems to indicate that they help provide security. Over-reliance on neighbors could apparently be a problem, however, as Cato later instructs the *vilicus* to limit the households with which he lends and borrows to only two or three (*Agr.* 5.3). Columella recommended avoiding having to borrow from neighbors because it would slow down the work (*Rust.* 11.1.20).[60]

3.7 Rural demand for coinage

Not everything could be acquired without the use of money or recourse to the market so there would also be rural demand for coinage.[61] As I have discussed elsewhere, Money Demand Theory "posits that money is just one of many forms in which wealth can be held and that an individual's demand for money will depend on the utility and potential returns offered by a range of assets."[62] The utility offered by cash translates into three different motives for holding it: transactions demand, speculative demand, and precautionary demand. Transactions demand refers to money held in order to make payments you know you will need to make with cash in the near future. The convenience of being able to pay immediately, in this case, outweighs the revenue lost from not having that money invested in some productive fashion. Speculative demand constitutes the money one holds in the expectation that the value of other possible investments will decrease.[63] Precautionary demand, finally, is "a cushion against unexpected need."[64] While rural demand for coinage was likely always less than urban demand, it was nonetheless considerable. As this chapter has shown, there were many things most farmers would have to buy. Farmers who wanted to continue practicing agriculture therefore had transactions demand to pay for such items. (Those in poor financial shape would have to go into debt and hope to be able to pay off their creditors after the harvest. I will discuss the sale of farm goods to raise money in the next chapter.) Cato recommends farmers keep speculative "balances" of wine and olive oil in order to wait for high prices (*Agr.* 3.2) and it is likely that savvy farmers did much the same with their money, hoarding coins to take advantage of low equipment prices or a nearby plot of land becoming available for purchase. Credit may also have been an option in both scenarios but interest rates would add to the expense. Finally, farmers would have had precautionary demand for coinage to pay for the unexpected maintenance expenses discussed above or the replacement of essential tools that were broken or stolen as well as structures destroyed by fire, flood, or marauding armies. But farmers would also have precautionary demand for money for reasons not directly related to agriculture. Willem Jongman has argued that "an important reason for rich Romans to hold large reserves in cash was the need to alleviate the complexities and unpredictability of property transfers from one generation to the next."[65] This insight can be expanded upon in at least one important way. Jongman's argument focused on the early Empire and the role of gold coinage in particular. He argued that "coin stocks ... were ... too large to be explained by the transactions motive" and suggested rich Romans kept large amounts of gold coins "to facilitate the execution of their wills, or to fully profit from what they received from the wills of others."[66] Justinian's *Digest* makes the usefulness of such reserves clear since it devotes considerable space to the "Action for Dividing an Inheritance." It quotes Ulpian to the effect that "If, in any action for dividing an inheritance ... the division is so difficult that it seems almost impossible, the judge can assign the whole condemnation and adjudicate all the assets to one party" (10.2.55).[67] In other words, it could be so difficult to divide up the assets of an estate among all the heirs that a judge might be forced to give all the property to one heir and order that individual to pay all the other heirs their shares in cash. The heir who gained possession of the property would have to sell at least some of the inheritance, perhaps under less than ideal circumstances, if he or she did not have ready money to give the co-heirs their shares. The practice of specifying cash legacies in wills further complicated these divisions. The second century jurist Gaius noted that, in actions on dividing an inheritance, "the judge's powers allow him to order that one or more items in the inheritance should be sold, and that the money raised from the sale should be paid to the man to whom it has been left as a legacy" (*Dig.* 10.2.56). Again, the forced sale of assets on short notice might not yield the goods' full value, diminishing the overall value of the inheritance. Heirs who had ready cash to pay off co-heirs would be well situated to receive the full value of their inheritances. While maximizing one's ability to fully profit from an inheritance might seem like a fairly poor reason to hold on to cash that might be profitably invested, it is worth recalling Cicero's claim that he had gained twenty million sesterces through inheritances (*Phil.* 2.40).

Another reason for the well-off to hold cash balances might be to ensure the establishment of endowments. L. Veturius Nepos of Feltria left HS 4,000 to the Herclanenses so that sacrifices could be offered at his tomb on the Parentalia. Another sum of money was given to a group of women to place flowers there (*CIL* V 2072).[68]

No doubt only the relatively wealthy had this particular kind of concern but a broader cross-section of the free population, urban and rural, is likely to have other death-related reasons for precautionary demand: fear of illness or injury requiring doctors, medicine, or offerings to the gods; and the desire to have one's body receive proper treatment after death. While the evidence is rather poor, there are ample grounds to believe that Romans spent considerable sums on doctors, drugs, and religious activities associated with health. Though he is a hostile and moralizing source, Pliny the Elder indicates that medicine was immensely profitable. He claims no profession was more lucrative and reports that imperial physicians could earn a quarter of a million sesterces a year (*HN* 29.2, 7). Only competition among greedy doctors, he claims, kept prices in check (*HN* 29.21). Cities offered various incentives to encourage doctors to set up shop within their walls.[69] Little, however, is known about their fee structures. The drug trade also thrived and Roman pharmacists drew upon a wide array of plants and minerals, some imported at great cost from beyond the borders of the empire.[70] Judging from Pliny the Elder's remarks, counterfeit drugs plagued the market. Those who could afford them no doubt sought out doctors while others purchased drugs, resorted to folk remedies, or appealed to the gods for help (as countless votive offerings attest). While the Italian countryside was perhaps not usually as unhealthy as urban areas, it is likely that all but the most impoverished households tried to keep some money in reserve to deal with medical emergencies. Healthcare-related precautionary demand was, in the aggregate, probably considerable.

Turning to death itself, although most of our funerary evidence comes from the outskirts of cities, there is no reason to believe that rural residents were less concerned about the proper treatment of their remains than their urban counterparts,[71] and it is quite clear that the funerals could be quite expensive. The coin sometimes placed in the tomb of the deceased might be the least of one's funeral expenses. One might hire *pollinctores* to wash and anoint the body, *vespillones* to carry it, *praeficae* to wail at the funeral, *ustores* to cremate the corpse, or *fossores* to dig a grave. Cremation required fuel and mourners often burned incense at funerals (see, e.g., Plut., *Sull.* 38). Prominent families might hire a *praeco* or herald to publicly announce the funeral, actors to impersonate their most famous ancestors and a *dissignator* to organize the procession. Musicians may also have played a role. While not providing any precise fees, two early imperial texts suggest that death brought the bereaved not just grief but also expenses. Pliny the Younger describes how his friend Fundanus had to divert money earmarked for his daughter's wedding to the purchase of incense, ointments, and perfumes for her funeral (*Ep.* 5.16.7). Martial (*Epigr.* 10.97.3) describes the preparations for a certain Numa's anticipated death which included the purchase of myrrh and cassia as well as the preparation of bier, pyre, grave, and the services of an undertaker (*pollinctor*). Ulpian says that funeral expenses "should be commensurate with [the deceased's] resources" (*Dig.* 11.7.14.6) and "are assessed according to the wealth or rank of the deceased" (*Dig.* 11.7.12). A cheap funeral for a wealthy individual might be construed as an insult (*Dig.* 11.7.14.10). Even though the deceased was not around to arrange his or her own funeral, it is clear that the estate of the deceased was generally expected to foot the bill. Ulpian declares that "it is best for the dead to be buried at their own expense" (*Dig.* 11.7.14.13). A woman's funeral was to be paid out of her dowry by whoever received it, her husband or father (*Dig.* 11.7.16). Furthermore, funeral expenses had priority over legacies as a claim on the estate (*Dig.* 11.7.14.1) and, according to the second century CE jurist Maecianus, they were to be paid even before debts owed by the estate (*Dig.* 11.7.45). Since the "dead [were] generally buried before anybody qualified as their heir" (*Dig.* 11.7.4), there could be considerable uncertainty about who should pay for the funeral in the meantime. Ulpian says that "anyone who spends something on a funeral is said to contract with the deceased, not with the heir" (*Dig.* 11.7.1), suggesting again the estate's financial responsibility, but it was not always so simple. It was understood that some people might arrange for the burial of a relative, friend, or loved one out of a sense of duty and *not* be entitled to recover the associated costs. There was a legal "action for funeral expenses" (*Dig.* 11.7.14.6). Ulpian explains:

> We can distinguish degrees of compassion; that is, the person who arranged the funeral may have been compassionate or dutiful to the extent of burying the deceased in order to prevent him from remaining unburied, but not to the extent of doing so at his own expense. If that is how it appears to the judge, he should not absolve the defendant. For who can bury a corpse which is somebody else's responsibility without in some measure feeling a sense of duty? So one ought to declare before witnesses whom one is burying and with what motives, so that there will not have to be an investigation later.
>
> (*Dig.* 11.7.14.7)

No matter how happy such declarations would have made the jurists, one may reasonably doubt their frequency. The amount of attention paid to the issue of funeral expenses by the jurists is a strong indication of the potential scale of those expenses and suggests in several ways that these expenses required someone to put up cash immediately, despite the risk of discovering that one was not in fact going to receive anything from the estate. We have much less information about the funerary expenses of non-elite Romans and the inhabitants of rural areas. Even if, as seems likely, they were nowhere near as expensive as elite funerals, they might still be expensive relative to the wealth of the deceased, as Ulpian's remarks suggest.

The cost of funerary monuments, it is important to emphasize, is a secondary concern. Though such structures could be costly, one could build them in advance, as some chose to, but, if one had not, there was no need for *immediate* expenditures since their design, the purchase (if necessary) of land on which to build them, and construction would take some time, during which financing could be arranged.[72] Unless a corpse was embalmed, however, the funeral had to take place on relatively short notice and could involve payments to various mortuary professionals with whom it might be difficult to arrange credit. It is likely that the funerals of rural dwellers would be less expensive than those of urban residents. Most would have had access to a family, hereditary, or patronal tomb (or simply land) in which to have themselves or their relatives buried. They would also have the tools to dig a grave or construct a bier, and probably even fuel for the pyre. While we have very little direct evidence for the costs of Roman farmers' funerals, Pliny the Elder gives us the obviously exceptional example of the freedman Gaius

Caecilius Isidorus who, in his will, claimed to own thirty-six hundred teams of oxen as well as more than a quarter million other head of cattle (not to mention sixty million sesterces in cash) and ordered 1.1 million sesterces to be spent on his funeral in 8 BCE (*HN* 33.135). The funerals of members of the imperial family could easily cost more. Nero supposedly burnt more than a year's supply of frankincense at Poppaea's funeral in 65 CE (Plin., *HN* 12.85). On the decurial level in the early Empire, Duncan-Jones suggested that two thousand sesterces was "a standard burial charge of long standing" since the sum appears repeatedly on funerary inscriptions.[73] Degrassi collected a broad range of funeral expenses for the early empire and came to similar conclusions for municipal elites.[74] At an even lower economic level, the Roman cemetery at Vagnari in Puglia shows that agricultural laborers working on an imperial estate might have grave goods, albeit of fairly low value. Among them were coins, glass vessels, iron nails, and even, in one instance, gold earrings. Even if these items were the possessions of the deceased or, in the case of ceramics, 'recycled' waste, they still speak to rural demand. Some of the graves also had funnels for offerings.[75]

3.8 Conclusion

Our sources – even the ones now most associated with the supposed ideology of self-sufficiency – overwhelmingly acknowledge and even encourage the purchase of some tools and supplies in the market. In fact, it appears very easy to spend lots of money in farming. Speaking of Sicilian agriculture, Cicero complained that "much annual toil and expense is applied to an uncertain end and result" (II *Verr.* 3.227). Much of the ostensibly self-sufficient Roman rhetoric is probably more an attempt to encourage the limiting of expenditures rather than their complete elimination.

In recent years there has been increasing interest in trying to quantify the Roman economy.[76] While we simply do not have enough evidence for the prices of farm equipment and supplies to quantify rural demand for them, it may be possible to work backwards from estimates of consumption to the *instrumentum* of the farm in order to get a sense of scale. If Rome was, as Rosenstein estimates, consuming around "160 million litres of wine" in the late second century, how many vinedresser's knives does that imply?[77]

The large-scale resettlement and colonization efforts of the late Republic and early Empire involved more than just the redistribution of land. Tiberius Gracchus, for example, used the money Rome inherited from the last king of Pergamum to provide equipment for those he was settling on *ager publicus*.[78] In only one instance, however, do we have a specific cost associated with a resettlement effort. In 180 BCE, the Romans deported forty thousand Ligurian men and their families to public land in Samnium. Even though the Ligurians brought some property with them to their new homes, the senate, Livy tells us, provided one hundred and fifty thousand sesterces "with which they could acquire what was needed for their new homes" (40.38). The numbers are suspiciously round but, if accurate, amounted to less than a *denarius* per head of household.[79] Judging from Cato's slightly later prices, this sum would not have been particularly generous. Even assuming the Ligurians brought a considerable amount of agricultural equipment, supplies, and livestock with them, building expenses could, as we have seen, be substantial. As Rosenstein has pointed out, "it is difficult to believe that the *patres* would have shown such solicitude when they voted to relocate the [Ligurians] and not regularly taken similar steps to ensure the success of new foundations of citizens and allies."[80]

Columella offers a different way to get at a sense of the scale of expenditure on farm tools. In his hypothetical calculations on the profitability of viticulture he suggests two thousand sesterces per *iugera* for the vineyard's 'dowry' of stakes (*pedamenta*) and withes (*vimines*), *twice* the estimated value of the land (*Rust.* 3.3.8). Given the poor reputation viticulture had for profitability, as Columella himself concedes, this is probably near the high end of the spectrum in terms of the ratio between the cost of equipment and the value of the land on which it was used.

Notes

1 Pliny (*HN* 18.41–3) relates the story of C. Furius Chresimus who, accused of using magic to steal his neighbors' crops, brought his slaves and excellently made iron tools (*ferramenta egregie facta*) into court and declared that *they* were his spells. Plutarch (*de Cup.* 2), admittedly writing in Greece, notes how people went into debt for Galatian mules and other non-essential purchases.

2 Pliny (*HN* 18.39) emphasizes that the most profitable way to farm was by minimizing expenses.

3 *Nundinae*: Andreau 1984; de Ligt 1991 and 1993, 107ff.; Frayn 1979 and 1993; Holleran 2012, Chapter 4; Ker 2010; Lo Cascio 2000; and MacMullen 1970.

4 de Ligt, whose work on the rural economy of the Roman Empire is essential, offers a list of "items of peasant expenditure" and helpful discussion (1990, 43).

5 Sharing: see Lirb 1993 for general discussion of cooperation among Roman farmers. Bowes et al. (2017, 186) suggest the "agro-processing point" they excavated at Case Nuove in southern Tuscany may have been "a collective endeavor, used by a variety of nearby farmers" for pressing olives and grapes.

6 The passage is difficult to interpret confidently. See Dalby (1998, 95) for some discussion.

7 Cistern construction: Thomas and Wilson 1994, 191.

8 Cement: in a discussion of cistern construction, for example, Pliny (*HN* 36.173) notes the need for clean sand, to which most farmers probably did not have ready access.

9 Rathbone (2008, 323) suggests that "most small Roman farmsteads were too flimsy and materially poor to leave much of an identifiable archaeological trace."

10 Harris (1993, 29) argues that "Pompeii makes it clear that the aggregate of metal artifacts ... to be found in an Italian city under the Flavians, was enormous." At the San Rocco villa at Francolise, nails were "found in layers of all dates" (Cotton and Métraux 1985, 153–6) while rather few were found at the nearby villa at Posto (Cotton 1979, 82).

11 See also Statius, *Theb.* 3.588–9 and Plin., *HN* 34.138.

12 Buried with tools: Small et al. 2007, 163 (possibly an early third century CE burial).
13 Regional traditions of cultivation: Varro, *Rust.* 1.50 and Horsfall 2001.
14 On the plow: White 1967, 47.
15 Dalby (1998, 191 n. 233) argues that Cato, *Agr.* 10 indicates that some Roman farmers used plows without a plowshare.
16 The mattock: White 1967, 38.
17 *Rastri:* used instead of a plow (Columella, *Rust.* 2.10.23); to dig out roots (Columella, *Rust.* 3.11.3); for *occatio* or breaking up the soil, i.e., harrowing (Plin., *HN* 18.180; Virg., *G.* 1.94–5); in the vineyard (Catullus 64.40). See also White 1967, 54–5.
18 Sharing plows: White 1970, 345.
19 White 1967, 44. *Sarcula* constitute almost half of the iron tools in a group found in the excavation of P. Fannius Synistor's villa at Boscoreale (Harvey 2010, 699).
20 White 1967, 50. The *marra* was another kind of hoe or mattock. White (1967, 42) suggests they were single-bladed and of medium weight. Although he concedes (1967, 41) that "neither the shape nor the function of this implement can be precisely determined," he argues that "[i]t must have been a common enough implement."
21 Types of *falces:* White 1967, 72.
22 *Falcula:* White 1967, 97.
23 *Falx messoria:* White 1967, 98.
24 Kron (2013, 219) suggests the "widespread introduction of professionally manufactured iron tools" helped bring about improvements in the productivity of Roman agriculture in the Empire.
25 Haarer (2013, 3499) notes that "deposits of iron ores tended to be fairly ubiquitous." There were placer deposits on Tyrrhenian beaches from Etruria to the Bay of Naples (Corretti and Benvenuti 2001, 136). Pliny (*HN* 34.142) says that iron was found nearly everywhere and was easily recognized.
26 Defining the *instrumentum* of a farm: Dig. 33.7.8.pr., 33.7.21, 33.7.25. Pliny the Younger may be referring in part to the cost of iron tools in a letter in which he discusses a neighboring farm he is interested in buying (*Ep.* 3.19). He notes that the current owner frequently sold his tenants *pignora* (security), reducing their debts but also their strength (*vires*).
27 See Bray (2010) for discussion of the value of iron in Roman Britain.
28 *Furcae:* White 1967, 107.
29 Survival of wooden tools: White 1967, 82.
30 Varro briefly discusses the purchase of oxen (*Rust.* 1.20.1), sheep (2.2), asses, and horses (2.6ff.), as well as providing the terms for purchasing cattle (2.5.10–11). Columella goes into more depth about the purchase of livestock: oxen (*Rust.* 6.2), cattle (6.20–1), sheep (7.2–3), goats (7.6), pigs (7.9), and even bees (9.8).
31 Mule-breeding: see Varro (*Rust.* 2.8) on the high prices fetched by breeding jacks.
32 Cato recommends buying keys and locks at Rome (*Agr.* 135.2).
33 Skins and barrels: Witcher 2016, 467. On the *culleus,* see Cato, *Agr.* 154.
34 See, for example, Cato, *Agr.* 10.4, 26, 54, 143.3.
35 Marzano (2007, 63) says the "production of bricks, tiles and pottery was a common commercial activity at both *villae rusticae* and *villae maritimae.*" Sometimes production "was aimed exclusively at the internal needs of the villa" (64). See, for example, *Dig.* 33.7.25.
36 Greene notes (1986, 158 and 164–5) that *terra sigillata* is "even found on relatively humble rural sites" and that "the scale of distribution of even the simplest wares suggests that they may have been sold at rural or even urban markets." He concludes that it "is quite clear ... that most Roman pottery was traded rather than made in households or estates for their own consumption."

37 Grain storage: Varro, *Rust.* 1.57 and 63; Columella, *Rust.* 1.pr.9–17.
38 On smallholders and agricultural equipment, see, for example, Witcher 2016, 467.
39 For example, Erdkamp who suggests (2005, 78–9) that "many peasants were by necessity underemployed." For further discussion see Chapter 4.5.
40 Seasonal labor demands of large, slave-run estates: see, for example, Garnsey and Saller 1987, 76–7.
41 Dalby (1998, 67 n. 27) interprets this somewhat curious recommendation as the desire of the owner himself "to make any contracts that will require more than a single day's work."
42 Free labor: Launaro 2015, 186.
43 Presses: Rossiter 1981, 348.
44 On the treatment of soil with lime, see: Kamprath and Smyth 2005, 357. Marzano (2015, 193) suggests that lime could not be produced near Rome so farmers there may have had to purchase it. Pliny (*HN* 17.42–8) discusses the use of marls but this practice seems mainly confined to Britain and Gaul.
45 Processing textiles: Flohr (2013b, 69–70) suggests large-scale fulling was done by intermediaries shortly prior to sale.
46 Uses of pitch: Cato, *Agr.* 2.3, 23, 25, 72, and 107; Plin., *HN* 14.127. For a pitch production facility in the Pyrenees, see Orengo et al. 2013.
47 Bitumen: Schwartz 2016.
48 Repairing cracked *dolia* with lead: Cato, *Agr.* 39.
49 Thurmond (2006, 234) notes there is much debate over how much salt a person needs. Modern pastured cattle get 35–45 grams/day but the amount varies depending on several factors (Donna Gatewood, personal communication).
50 Salt for stored fodder: Cato, *Agr.* 54.2. To preserve olives: Cato, *Agr.* 23; Columella, *Rust.* 12.49.6, 12.50.3. To make brine: Columella, *Rust.* 12.6.
51 Clarysse 2013, 6022. The *Digest* refers to a case in which a man left the usufruct of salt pans to his wife but the location is not specified, nor are the pans necessarily associated with the farms also mentioned in the passage (33.2.32.2–3).
52 See Giovannini (1985) for further discussion of the role of salt pans in early Roman expansion, and Plin., *HN* 31.89.
53 Sea salt: Thurmond 2006, 239.
54 Crude salt: Thurmond 2006, 246.
55 Columella also wrote a non-extant *Adversus astrologos,* criticizing Chaldaeans (*Rust.* 11.1.31).
56 Pliny also relays the recommendation to sow navews while naked and praying (*HN* 18.131).
57 Rural festivals: Stek 2010, 200.
58 Coins and festivals: Stek 2010, 176–8.
59 Parentalia: Dolansky 2011, 125–57.
60 Neighbors: Columella indicates that there was some debate over the importance of the quality of neighbors in determining whether to buy a particular property (*Rust.* 1.pr.5).
61 For an earlier discussion of this issue with respect to the late Republic, see Hollander 2007, 122–35.
62 Money Demand Theory: Hollander 2007, 141–2.
63 Speculative Demand: Serletis 2001, 60.
64 Precautionary Demand: Mishkin 1992, 532.
65 Intergenerational property transfers: Jongman 2003b, 191.
66 Gold coins and facilitating the execution of wills: Jongman 2003b, 191 and 195.
67 Translations of the *Digest* are from Watson 1998.
68 See also the will of Q. Cominius Abascantus of Misenum: D'Arms 2000.
69 Incentives for doctors: Scarborough 1969, 111–12; King 2001, 34.
70 The drug trade: Nutton 1985, 138–45; Jackson 2005, 216–17; and King 2001, 41.

71 Small et al. (2007, 125) note that there is considerable variation within Italy in burial customs.
72 The jurist Macer notes that "funeral expenses are generally held to include any expenditure on the body (for example, on ointments), the cost of the place where the deceased is buried, and any rent which has to be paid, or the cost of the sarcophagus and the cost of transporting the body" (*Dig.* 11.7.37). He also refers to a rescript of Hadrian to the effect that the cost of construction of a monument is not to be considered a funeral expense.
73 Burial charges: Duncan-Jones 1962, 62 n. 43.
74 Funeral expenses of municipal elites: Degrassi 1960, 236.
75 Small et al. 2007. Finds from other rural grave sites reveal similar practices, for example: Mercando 1965; Sena Chiesa 1985; and Passi Pitcher 1987.
76 See especially the volumes in the Oxford Studies on the Roman Economy series.
77 Wine consumption estimate: Rosenstein 2008, 5.
78 Plut., *T. Gracc.* 14. On this episode, Rosenstein observes (2004, 166) that the "attraction of Gracchus's proposal and other land laws that followed ... may have lain as much in the fact that they sought to create working farms by providing the equipment and livestock necessary to make a smallholding self-sufficient as because they made allotments available in the first place."
79 On suspicious numbers: Scheidel 1996.
80 Relocation of the Ligurians: Rosenstein 2004, 166.

4 Vendors and lenders

The rural supply of goods and services

4.1 Introduction

Cato the Elder begins the *De agricultura* by distinguishing between three ways of making money: trade, moneylending, and farming. He describes agriculture as less risky than trade and much more respectable than moneylending but nevertheless a source of profit. In addition to providing strong, brave men, agriculture provided *quaestus stabilissimus*, the most stable profit (*Agr.* pr.4). However, Plutarch, in his biography of Cato, says that he eventually came to consider agriculture more of a pastime than source of income (*Cat. Mai.* 21). Instead, he invested his resources in certain and safe enterprises such as lakes, hot springs, *fullonica*, pitch works, pastures, and forests. These provided great profit but could not, Cato said, be damaged by Jupiter.

This question of the profitability of agriculture is a persistent one, both in ancient Rome and more recently.[1] In his Verrine orations, Cicero, presumably trying to win sympathy for his Sicilian clients, says the reason for farming is pleasure and the hope of more rather than profit (*Verr.* II 3.227). He emphasizes the uncertainty, how one is at the mercy of the weather and market prices rather than one's knowledge and effort. But elsewhere, Cicero presents a more optimistic view. In the *De officiis*, he has Cato the Elder claim that nothing is more profitable than agriculture (1.151) and in the *De senectute* he likens agriculture to a bank which never rejects an order nor returns what was invested without interest, although he admits that the interest rate varies (*Sen.* 51). Later writers provided more nuanced and coherent views. I have already mentioned Columella's analysis of the profitability of viticulture (*Rust.* 3.3). Pliny emphasizes the prudent investment of labor, quoting an old saying that nothing is less profitable than really good farming (*HN* 18.36), which he later calls "destructive" (*HN* 18.38). The reason for this appears mainly to be the cost of the labor (*impendium operae*); Pliny highlights the difficulty of olive cultivation and (curiously) Sicilian agriculture in particular. Given the variables of soil, crops, climate, weather, market, resources, and experience, it is perhaps pointless to attempt to generalize about profitability. Our sources make clear that some farmers succeeded while others went bankrupt.

A major cause of failure was the many expenditures, necessary or unnecessary, that farmers made. Having discussed in the previous chapter the goods and services which farmers could not acquire locally or produce for themselves and, indeed, that they likely wanted coinage *per se*, in this chapter I turn to the question of how they would get the money they needed and wanted. There were four main means of acquisition: the sale of one's produce, the sale of one's labor, the sale of other assets, and moneylending (by which I mean both earning money through loans and borrowing money). We cannot forget, however, that much wealth was transferred to the countryside in an extra-mercantile fashion by the wealthy investing profits gained elsewhere in rural properties for the purpose of leisure; veterans flush with plunder settling down on land assigned them, or purchased, upon discharge; and perhaps even merchants seeking a more respectable occupation (see Cic., *Off.* 1.151; Petron., *Sat.* 76). All these things, but particularly the first two, happened quite frequently in the late Republic and early Empire. I will return to this issue in the following chapters but here it is enough to acknowledge that much wealth was transferred to the rural economy outside of the market.

In addition to surveying the various means of money acquisition, with particular attention paid to farm produce, I will consider their relative importance and the factors which would have altered their attractiveness in the late Republic and early Empire. Obviously, anything of value that a farm might produce could at least potentially be sold for money so it is impossible to be exhaustive, but it will become clear that Roman farmers had several plausible strategies to earn cash. The relative importance of different agricultural strategies for generating income in the form of coinage would have varied, of course, from place to place and over time. The factors mentioned in Chapter 2 (soil, climate, access to irrigation, availability of labor) are of fundamental importance but the nature of demand and proximity to markets were also crucial issues.[2] We can get some sense of the nature of the market for agricultural products by looking at the evidence for prices, profits, the geographical extent of trade, the role of these foods in the Roman diet, in medicine, the other uses of these products (and associated byproducts), and also considering the constraints and synergies they might entail in terms of production, transportation, and trade, by which I mean the requirements of time, capital, and equipment for their production and processing.

4.2 Animals and animal byproducts

In the *De officiis* (2.89), Cicero tells the story of a conversation Cato the Elder once had with an unnamed interlocutor. Asked what was the most profitable use of one's property, Cato responded 'raising animals well' (*bene pascere*). What was next? 'Raising them reasonably well' (*satis bene*). Third best? 'Raising them poorly' (*male pascere*). Arable farming (*arare*) came in fourth and Cato likened moneylending to murder. At least in Cato's period, it seems clear that raising livestock was the most profitable aspect of agriculture. There were good reasons for this and it is likely that pastoralism (broadly defined) remained quite lucrative throughout our period. There would usually be considerable urban demand for cattle to sacrifice but it was the versatility of the major kinds of livestock that was especially advantageous to the farmer. In addition to meat, some farm animals could provide milk, cheese, hides, and/or wool. Cartilage could be made into glue; horncones and footbones could be used in tanning.[3] Cobblers used the sinews as thread while the army needed them for artillery.[4] Even bone ash had applications in glassmaking.[5] Some animals provided transport or motive force for plowing (and their services could be hired by others for additional income) but all provided manure, which, of course, greatly enhanced arable agriculture. These animals could also, in some cases, graze on marginal land or consume farm byproducts. As Foxhall notes, sheep, goats, donkeys, pigs, and cattle can eat "leaves of pruned branches of all fruit trees ... the press cake from wine and oil processing, and fruits spoiled or damaged by pests and diseases."[6] We should not be surprised that, as Kron points out, in convertible husbandry, most of the profit tends to come from the livestock.[7]

4.2.1 Meat

A major reason for livestock's profitability was Roman demand for meat. Zooarchaeology continues to greatly enhance our understanding of Roman meat consumption (and indeed animal husbandry in general), although predictable gaps remain.[8] The faunal evidence indicates that, while there were many variations in consumption – between city and country, between northern and southern Italy, as well as across time (seasonally and from Republic to Empire), by class, and individual taste – the Romans consumed a large, if unquantifiable, amount of meat.[9] This consumption, primarily of pigs, cattle, and to a lesser extent sheep and goats, increased starting in the third century BCE.[10] Thanks to breeding efforts and improved agricultural knowledge, the size of Roman livestock also increased.[11]

How much demand for livestock was there for Roman farmers to exploit? To a greater extent than in rural areas, urban dwellers seem to have consumed domesticated animals as opposed to wild game.[12] Unfortunately, we lack the sources to provide a detailed view of the urban meat trade.[13] It seems likely that animals brought in from the countryside were purchased by butchers who then sold the meat in shops or that the livestock was sold for sacrifice and then leftover meat was sold to butchers.[14] Holleran has suggested, for example, that the Forum Boarium at Rome probably served "more of a wholesale function for butchers and perhaps to a lesser extent, priests or those charged with obtaining beasts for sacrifice."[15] She notes that, due to its processing and storage requirements, meat was "more likely to be sold in fixed locations."[16]

The consensus now seems to be that nearly everyone would have been able to afford meat although poorer Romans would have fewer and less appealing choices.[17] The owners of large herds who focused on the sale of meat and animal byproducts would sell their animals at a relatively young age. The cattle found on rural sites tend to be older, perhaps typically slaughtered when no longer useful for farm work.[18] Smallholders probably ate some meat but were unlikely to regularly slaughter larger livestock since even a single sheep would provide over 40 pounds of meat.[19] Cattle were apparently more common in northern Italy while sheep and goats predominated in the south and "forested districts" throughout the peninsula had more pigs.[20] Varro implies that most farms would have at least a few pigs, since he has Scrofa declare that their fathers regularly called "lazy and wasteful" anyone who bought pork from a butcher instead of from their own farm (*Rust.* 2.4.3). One advantage of pig production, it has been suggested, was that pigs were in less direct competition with humans for resources.[21] Kron, however, has argued that it is wrong to think in terms of livestock competing with humans for resources when, thanks to the practice of convertible husbandry, the presence of the livestock dramatically increases agricultural yields.[22]

The Roman army also likely had a substantial demand for livestock. Meat was a regular part of the diet of Roman soldiers,[23] and hides were needed for military equipment. One estimate suggests that sixty-eight thousand goatskins were needed to produce tents for a single legion, and three thousand cattle hides for their boots.[24]

There were, of course, many meat products available beyond pork and beef. On the one hand, there was fowl, particularly chicken – apparently a fairly common farm animal. On the other hand, there was a variety of more exotic fare, the results of *pastio villatica*. In Book 3 of his *De re rustica*, Varro has Merula explain that there are three kinds of *pastio villatica*, involving *ornithones* (aviaries), *leporaria* (hare warrens), and *piscinae* (fishponds) (*Rust.* 3.3.1). He clarifies that *leporaria* were not just for hares but also boar, deer, bees, snails, and dormice. *Ornithones* could be for a variety of birds (including peafowl, fieldfares, geese, and ducks). *Piscinae* could be fresh- or saltwater fishponds. Varro indicates that *pastio villatica* could be quite lucrative (*Rust.* 2.pr.5) but many forms of it were beyond the reach of most farmers. Wild game preserves required much more land than smallholders were likely to own or have access to and saltwater pisciculture required particular kinds of land, significant capital outlays, and expertise.[25] Varro does, however, say that the *plebs* raised freshwater fish "not without profit" (*Rust.* 3.17.2). The raising of smaller animals like dormice could also be accessible to smallholders but unfortunately Varro spends little time on them. For the vast majority of farmers, however, most pisciculture and many forms of *pastio villatica* were probably not an option.

Starting in the second century BCE there were attempts to regulate meat consumption, some aspects of which were deemed problematic, as part of sumptuary legislation. The *lex Fannia* of 161 BCE, for example, limited the consumption of poultry at dinners as well as overall expenses.[26] The sumptuary laws do not seem to have had much long-term impact on Roman demand for meat. In the late Republic, Varro indicates that there was a great deal of money to be made from poultry on the urban market and he devotes about half of Book 3 of his *De re rustica* to *ornithones*. The character Appius says that, with the help of a banquet and a triumph, five thousand birds could earn the owner sixty thousand sesterces. Interestingly, he goes on to suggest that the profits be lent at interest (*Rust.* 3.5.8). Another character, Merula, later reports that Marcus Aufidius Lurco used to earn the same amount each year from his peafowl (*Rust.* 3.6.1). These are, of course, conventional large numbers but they do indicate the potential and scale of aviaries oriented towards the demand of the city of Rome. Axius (another character in the dialogue) claims that peafowl are more profitable than chickens (*Rust.* 3.4.1). A variety of other birds were sold, including, according to Varro, ortolans, quail, turtle-doves, geese, and ducks. Varro (*Rust.* 3.7.10–11) relates the story of a *mercator* haggling with an equestrian over the price of squabs (young pigeons). However, he seems to indicate both that the equipment necessary for raising pigeons could be quite expensive and that it was practiced in urban areas (*Rust.* 3.7.11). The farmer might sell eggs too (Columella, *Rust.* 8.5.4 and 8.6.2). Roman demand for poultry remained strong under the Empire. Columella reports good prices for chickens raised near the city (*Rust.* 8.5.9) while Pliny notes the recent introduction of new birds (Plin., *HN* 10.69).

4.2.2 Wool

The sale of wool would be an option of variable appeal to farmers. On the one hand, it is clear that there was sizable urban and rural demand for woolen goods and that, in the aggregate, woolen textiles were worth a considerable amount of money.[27] However, our sources indicate that different kinds of sheep produced wool of varying quality and some regions were better suited to raising them than others. Jongman notes that "good quality wool" only came from a few areas in northern and southern Italy.[28] He suggests that, after "initial cleaning" (which substantially reduced the weight), raw wool was exported to "urban population centers in central Italy" for further processing into textiles.[29] Some estates opted for full domestic production, as we know from Varro's remarks (*Rust.* 1.16.4) about having fullers on an estate's staff. Given the constraints on the labor of smallholders, it is doubtful they would have been able to produce much wool for the market. As I noted in Chapter 2, they were likely to be net consumers of woolen textiles.

As our sources indicate, spinning and weaving were standard domestic tasks for women. Flohr calls this the "traditional model of self-sufficiency, which pervaded (female) elite ideology."[30] No doubt these were common household tasks for women across all economic strata,[31] but smaller households likely had to supplement domestic production (if any) while upper class households probably sought at least some 'luxury' textiles from the market. In one of his diatribes against luxury, Columella complains that the women of his time no longer oversee wool-working, do not like home-made clothes, and spend outrageous amounts of money on their dress (*Rust.* 12.pr.9). As Flohr also reminds us, we must also consider how textiles were maintained and reused.[32] Old clothes could be repaired but also sold.

4.2.3 Dairy products

What about Roman demand for milk? It used to be thought that the Romans did not drink much milk. Columella and others characterize *nomads* as milk-drinkers (*Rust.* 7.2.2), but this is more literary trope than reality. While many in southern Italy were lactose intolerant,[33] there is still plenty of evidence for Roman milk consumption.[34] It is now thought that it was not unusual to drink milk, especially in the countryside.[35] Varro says milk is the most nourishing drink (*Rust.* 2.11.1). Pliny the Elder says goat's milk is the most nourishing (after breast milk), but that the most efficacious (*efficacissimum*) comes from asses, and sheep's milk is "sweeter and nourishes more" than cow's (*HN* 28.124). Galen calls *good* milk the most wholesome food you can consume (*De alimentorum facultatibus* 3.14), and his writings imply that doctors routinely had to deal with the effects on their patients of milk consumption.[36] Pliny recommends milk for a variety of ailments.[37] He reports that Nero's wife Poppaea had a flock of five hundred she-asses so she could bathe in ass's milk wherever she went, to whiten her skin and "believing it smoothed the skin" (*HN* 11.238).

Because it would spoil quickly,[38] milk could only be sold in cities if the herds producing it were located nearby. There is some indication that farmers regularly took advantage of their proximity to urban areas in order to sell milk. Varro relates the story of a goatherd who earned a *denarius* per goat per day at Rome (*Rust.* 2.4.10). Virgil mentions shepherds taking milk (and cheese) to town for sale (*G.* 3.400–3). Columella advises quickly selling your lambs to the butcher, if you live near town, because you can profit from the mother's milk (*Rust.* 7.3.13). Calpurnius Siculus, finally, refers in passing to the urban sale of milk by a "not quiet" shepherd (*Ecl.* 4.25–6), perhaps indicating that shepherds tended to loudly roam the city in search of customers.

With respect to cheese production, the evidence is poor. There are a few mosaics and sarcophagi with milking scenes, assorted cheese molds, strainers, and carbonized cheeses from Herculaneum as well as some depictions in frescos. It does not amount to much. As with milk, we must rely largely on the scattered textual evidence.

Cheese was a basic part of the diet of country folk (Columella, *Rust.* 7.2.1) and Roman soldiers,[39] and but the elite enjoyed it too. Columella says that sheep cheeses appeared on "the tables of refined people" (*Rust.* 7.2.1). According to Suetonius, Augustus particularly enjoyed "moist, hand-pressed cheese" (*Aug.* 76), while Galen notes that wealthy Romans enjoyed a kind of cheese called *bathysikos* (*De alimentorum facultatibus* 3.16). Pliny, predictably, mentions some cheese-based remedies (*HN* 28.131–2). Cheese (along with honey) features regularly in Cato the Elder's recipes,[40] and seems to have been relatively inexpensive.[41]

Columella advises making cheese when one is far away from markets; when close, it made better sense to bring the milk to town in pails or sell soft cheese quickly (*Rust.* 7.8.1). Harder cheeses could be exported *trans maria* (*Rust.* 7.8.6). The trade in hard cheeses was extensive, and a wide variety were available in Rome: from the villages of Gévaudan near Nemausus (but only when fresh); Docleate cheese from the Dalmatian mountains; Vatusic cheese from the Alps; Apennine cheeses (e.g., Coebanum, a sheep's cheese, from Liguria); Sarsina cheese from Umbria; giant cheeses from Luna; Vestinian cheese produced near Rome; and even some from as far away as Bithynia (Plin., *HN* 11.240– 1). Rome itself produced an excellent smoked cheese (Mart. 11.52.10).

Cheese-making was a seasonal activity from May to mid-July (Varro, *Rust.* 2.11.3–4). Transhumance might mean that peak cheese-production times were spent in the mountains and shepherds would return to the lowlands in the fall laden with cheeses.[42] To make cheese one needs rennet (though there were plenty of readily available animal and plant-based sources),[43] as well as salt (Cato, *Agr.* 88), containers, presses, and fuel for processing the milk. Undoubtedly, the livestock itself was the biggest capital investment. One nice synergy of cheese production is that whey, the cheese-making byproduct, could be fed to pigs (Cato, *Agr.* 150) and "imparts a wonderful sweetness" to the meat.[44] Cato indicates that some Romans leased their flocks to shepherds and got rental payments in kind in milk and cheese (*Agr.* 150). There were important tradeoffs with respect to different kinds of livestock and cheese-making. Pliny says cow's milk makes twice as much cheese from the same amount of liquid as goat's milk (*HN* 11.238). According to White, "goat's milk has a higher yield than sheep's milk and was preferred for ... cheese-making,"[45] but most cheese probably was made from sheep's milk in peninsular Italy.[46] Production of hard cheese would be preferred because it was more difficult to control the temperature and humidity when making soft cheeses.[47] Garnsey notes that the keeping of sheep was "standard practice" on the Roman farm,[48] and Frayn speculates that subsistence farmers used milk to make cheese on a scale that exceeded that of meat and wool production.[49] Kron suggests that the cheese 'market' may have rivaled or exceeded that of *garum* and salt fish.[50]

As for butter, Pliny implies it was not especially important among the Romans (*HN* 28.133–4). He does, however, say it was used to anoint Roman children (*HN* 11.239) and was an ingredient in some drugs.[51]

4.2.4 Apiculture

The production of honey was another potentially lucrative agricultural sideline, and one to which many farmers could have had recourse. Our non-literary sources are once again rather poor. No Roman hives have turned up in excavations so far (probably because Roman beekeepers preferred hives made out of perishable materials rather than ceramics). There are no known illustrations of Roman hives or beekeeping.[52] One possible bee-like creature does appear on the Eumachia frieze from Pompeii,[53] and bees serve as countermarks on a few Roman coins.[54] We must again rely mainly on the textual evidence, which is refreshingly substantial (Varro's discussion of the topic is exceeded in length only by that of aviaries and Columella also devotes much space to them).

While there are some hints that honey was a delicacy, the evidence overall suggests it was in fairly widespread use. On the one hand, *mulsum* (honey-wine) may have been a special treat. Q. Terentius Culleo served it to guests at the funeral of Scipio (Livy 38.55) and Quintus Cominius Abascantus left HS 110,000 to the decurions of Misenum so that such wine could be served to them and the people on his birthday.[55] Varro also implies that *mulsum* was expensive due to its honey (*Rust.* 3.16.1). On the other hand, surviving recipes and descriptions of meals suggest widespread use of honey in cooking.[56] Galen says that both country-folk and poor urban residents made flat cakes and then dipped them in warm honey (*De alimentorum facultatibus* 1.3).[57]

Honey was also an ingredient in all sorts of medications: pennyroyal mixed with honey and soda was thought to cure intestinal problems (Plin., *HN* 20.153); honey mixed with pounded horehound was good for certain male genital afflictions (*HN* 20.243); honey mixed with *costum* helped women's skin and, if mixed with aloe, helped with bruises (*HN* 21.76); and white honey was used for ulcers and certain eye problems (*HN* 11.34.38). According to Pliny (*HN* 22.114), a certain Pollio Romilius attributed his vigor after more than 100 years of life to the drinking of honey wine. Honey in which bees had died could also be an ingredient in drugs: for dysentery Pliny notes the use of blackbirds roasted in such honey (*HN* 30.58). In addition, honey could be used to embalm people, as a poison (the Heptacomitae used 'bad' honey to drug Pompey's soldiers prior to an attack, Strabo 12.3), or to hide poison (Plin., *HN* 21.78). More common uses would have been as a food preservative,[58] in the preparation of perfumes and cosmetics,[59] and as a libation.

We lack much price data for honey but it was at least *relatively* expensive.[60] Columella warns against robbers of honey (*Rust.* 9.6.4) and there are certainly indications that the sale of honey could be lucrative. Pliny the Elder says that bee keeping was particularly profitable (*HN* 21.70). Varro reports that a certain Seius rented his Ostian apiaries for five thousand pounds of honey per year (*Rust.* 3.16.10). He also records that the Veianii brothers developed a dedicated apiary on a very small villa near Falerii and earned at least ten thousand sesterces per year from it (*Rust.* 3.16.11). The freedman *mellarius* Aulus Fuficius Zethus, perhaps a honey dealer, did well enough to have a funerary inscription set up for him at the Porta Trigemina in Rome (*CIL* VI 9618 = *ILS* 7497). Roman writers appear to be connoisseurs of honey, familiar with varieties coming from all over the Empire.[61]

Apicultural byproducts could also be lucrative. Varro comments that *propolis* (bee-glue), used to make poultices, was more expensive than honey on the Via Sacra (*Rust.* 3.16.23). Even beeswax was an object of trade. According to Pliny, wax had a thousand uses (*HN* 11.11). There were wax candles, tablets, masks, and wax was used in the casting of some metals. Wax had a role in medicine too, served in a broth, for example, to those suffering from dysentery (*HN* 22.116–17). Wax came to Rome from all over the empire, often in the form of tax revenue. The Corsicans paid two hundred thousand pounds of beeswax to Rome after being defeated by the Praetor C. Cicereius in 173 BCE (Livy 42.7.1). The Sanni (from Pontus) also sent wax as tribute (but not their honey, which was poisonous) (Plin., *HN* 21.77). Pliny mentions Cyprian, Punic, Pontic, and Cretan wax among others (*HN* 20.240, 21.83). Curiously, despite its many uses, Columella says that wax was of little value (*Rust.* 9.16.1). Perhaps the wax imported as tax revenue depressed prices on the Roman market.

Pliny did not consider beekeeping to be expensive (*HN* 21.70) and echoes Varro's remarks that the best hives were made from bark (*Rust.* 3.16.17). Many of the recommended materials for hive construction were presumably readily available in the Mediterranean basin: giant fennel, withies (willow shoots used in basketry), cork-trees, hollow trees, and hives constructed from wooden boards (Plin., *HN* 21.80; Varro, *Rust.* 3.16.15; Columella, *Rust.* 9.6.1), even earthenware, although our sources strongly advise against its use because of its poor thermal qualities and fragility (Varro, *Rust.* 3.16.17; Columella, *Rust.* 9.6.2). Columella cautions against brick and dung-based hives because they would be difficult to move and were, in the latter case, susceptible to fire (*Rust.* 9.6.2).

You could buy swarms, but it was also possible to capture them (Columella, *Rust.* 9.8.1). Varro provides advice on buying bees (*Rust.* 3.16.20–2), and Columella implies long-distance trade in swarms because he warns of the dangers of changes in climate (*Rust.* 9.8.3). Petronius has Trimalchio claim to have brought bees from Athens to Italy (*Sat.* 38).

With respect to labor requirements, Columella does say that one's bees always need attention (*Rust.* 9.9.1) but his focus is large-scale operations. He recommends there be a building adjacent to the hives for the beekeeper to live in and for equipment storage (*Rust.* 9.5.3). Other sources suggest that beekeeping was not necessarily a full-time job. After the winter, one needed to open, inspect, and clean out the hives (Columella, *Rust.* 9.14.1). There was much work to be done during the harvest: smoking the hives, cutting out honeycomb, extracting the honey, and processing the honey and wax.[62] The first harvest was in spring,[63] a second, providing the best honey, was in late summer (Varro, *Rust.* 3.16.34; Columella, *Rust.* 9.14.10). There might be a third harvest in the fall but some sources advise against it and it apparently produced the worst honey. Varro recommends checking the hives three times a month during the spring and summer to see if cleaning is necessary (*Rust.* 3.16.17). Caterpillars and moths could cause trouble

(Columella, *Rust.* 9.14.5). Over the winter or in bad weather, it might be necessary to provide food for your bees (Varro, *Rust.* 3.16.28). Compared to the daily attention required by most large kinds of livestock, bees would not have been arduous to keep. Since honey lasts a very long time in storage, it would be possible to wait for ideal market conditions to sell it (Varro, *Rust.* 3.16.11), a definite advantage over meat and cheese. The necessary equipment included knives (Columella, *Rust.* 9.15.4), fuel, and containers for collecting and storing the honey, for washing and melting the honeycomb fragments into wax, as well as molds into which to pour the wax (Columella, *Rust.* 9.16.1). These probably did not entail a major capital investment.

4.3 The profitability of plants

Cicero has Cato the Elder paint a grim picture of the relative profitability of arable farming but was the situation as poor as he implies? For the 'Mediterranean Triad' of grain, olives, and grapes, the prospects of a good return were not great for the average farmer (although much depended on location, especially proximity to transportation and markets) and no crop seems to have offered the versatility of, for example, sheep (offering milk, cheese, meat, wool, hides, and manure). Nevertheless, a variety of vegetal products did offer the chance of reasonable returns and to a broad spectrum of farmers, not just the wealthiest.

4.3.1 Grain

To what extent was grain a crop which could earn a Roman farmer money? Our limited sources do not say much about the sale of grain by peasants or otherwise.[64] Erdkamp suggests that small farmers probably "sold in bulk to local millers and bakers, and to traders."[65] He points out that even the sale of hay could be profitable since cities would need fodder for urban work animals.[66] However, one major reason to think that the sale of grain was *not* a major item of trade for farmers, at least in the vicinity of Rome, is the large amount of grain flowing into the city as tax revenue from the provinces. This artificially elevated supply would have depressed prices considering that, in the late Republic and early Empire, much of it was sold at a subsidized price or distributed free to those eligible. Grain would not have been an especially good choice for a farmer as a reliable source of profit under those conditions. In fact, Varro complains that greed had led the Romans to make meadows out of fields of grain (*Rust.* 2.pr.4). Suetonius indicates that Augustus was aware of this problem. He writes that the emperor had contemplated ending the *frumentationes* (grain distributions) and, although he ultimately decided against this, he nevertheless subsequently managed the grain supply (*annona*) in such a way as to take into account the interests of farmers and traders (*Aug.* 42). Yields, if they were as low as Columella says they were (4:1) in most parts of Italy (*Rust.* 3.3.4), would certainly not have encouraged a focus on grain.[67] In other parts of Italy, beyond the vicinity of Rome and its transportation network, grain may well have been somewhat more appealing but other products offered greater potential.[68]

4.3.2 Viticulture

For those with the capital, land, and knowledge, viticulture could be a source of profit, although Varro and Columella's remarks indicate that results varied and that it was not difficult to lose money producing wine. Ceramic evidence gives us a partial view of the Roman wine trade and shows, for example, the widespread distribution of Italian wine amphorae in Gaul in the late Republic. However, wine (and oil) could also be stored and transported in skins, and barrels came into use for wine in the early Empire as well. Much of the wine produced in the neighborhood of Rome likely reached the urban market without the aid of ceramic containers and so its scale is that much more difficult to estimate.[69]

Given the Roman diet (Rosenstein estimates wine consumption at one hundred liters per person per year),[70] wine would always have been in considerable demand. The wines of some regions were famous and so presumably more lucrative than those produced elsewhere. There is also clear evidence for large-scale exports (i.e., to Gaul and the eastern Mediterranean).[71] These were probably also profitable ventures considering how long they lasted. But what of smaller scale operations in less favored locales? We must imagine that a wide spectrum of wines of varying quality and price were usually on the market and that a range of producers of varying sizes sold wine. It is not clear what proportion of wine production was of high quality and so could command high prices. Carandini estimated that such wines accounted for only 20 percent or less of overall production.[72] Ordinary wines that could not be sold at a premium would not be especially lucrative. Jongman and Rosenstein have, furthermore, raised important doubts about the potential profitability of wine production. Jongman points out that relatively little land was required to produce enough wine (and olive oil) to meet Rome's demand,[73] while Rosenstein has cast doubt on the scale of demand.[74] This meant most winegrowers had little chance of substantial profits.

Even though margins may not have been great, it is clear that there was considerable local, regional, and long-distance trade in wine. Some farmers sold their wine to merchants "at the gate," even in advance of the harvest,[75] while others might transport it and market directly to consumers. There is ample evidence for intermediaries in the wine trade.[76] Varro notes (*Rust.* 1.16.2) that many farms had to import wine. Perhaps some of it was purchased from other farms nearby that were more favorably situated for viticulture. Also, we should not neglect to consider the other potentially lucrative products of viticulture: table grapes, raisins, cuttings, and grape syrup. Thurmond argues that "grape syrups were the primary sweeteners of Roman antiquity, at least among the poor."[77] The extent to which wine offered a Roman farmer a source of revenue would depend, as it did for most agricultural products, on a host of factors including quality of land, proximity to markets and transportation, and availability of labor.

4.3.3 *Olive oil*

Olives are at least a potential cash crop for smallholders but Erdkamp has argued that they were "unattractive" to such farmers because of the considerable labor demands.[78] Wealthier farmers who could marshal enough domestic labor and afford seasonal workers were more likely to produce for the market. Cato mentions the sale of olives on the tree and this must often have been an appealing option for larger producers seeking stable revenue.[79] Aggregate demand for olive oil was clearly substantial. Garnsey estimates food consumption at twenty liters per person per year, while Hitchner suggests twenty to twenty-five liters.[80] Hitchner further estimates Rome's demand to be more than twenty-five million liters per year.[81] The use of olive oil as a fuel and in bathing obviously contributed to the level of demand. (Table olives could also be sold.) Mattingly argues that it had "enormous potential for marketability."[82] Some doubt, however, that the trade in olive oil matched the scale of the wine trade.[83] As with grain, the fact that the state seems to have artificially stabilized the price of olive oil (Plin., *HN* 15.2) may have dissuaded those close to Rome from employing it as a cash crop. It is important to keep in mind that, as with wine, there were gradations in the quality of olive oil. Columella notes the great value of green oil compared to other kinds (*Rust.* 12.52.2 and 20). As with wine, olive oil's potential for profit would have varied considerably.

4.3.4 *Flowers*

Flowers could provide a significant profit for a farmer close to the right market. The reason for this was the major role of flowers in Roman religion. Most Roman religious rites involved flowers to one extent or another, including widely practiced observances such as the Parentalia and, of course, burial rites.[84] Flowers might adorn statues and be purchased near sanctuaries. Fronto, for example, mentions *coronarii* selling flowers and wreaths at the temple of Portunus (*Epist.* 1.7.2). Cato recognizes the potential of flowers, recommending their cultivation in gardens near town (*Agr.* 8.2). Varro makes the same point with respect to violets and roses (*Rust.* 1.16.3), while Cicero seems to allude to the profitability of the flowers produced by his Tusculan estate (*Fam.* 16.20). We lack, unfortunately, the sort of evidence to indicate the importance of the flower trade relative to other plant products.

4.3.5 *Vegetables*

Vegetables also held great promise as a source of cash. Again, proximity to urban markets would have been critical. We hear of gardeners selling vegetables in town (and of the wealthy buying them despite idealizing domestic production). As Holleran points out, fresh vegetables and fruit, since they did not require preparation but would spoil quickly, were well suited to street traders.[85] At Rome, the Forum Holitorium also served the needs of vegetable sellers and their customers (Varro, *LL* 55.146). Simulus, an admittedly problematic fictional example, sells asparagus, gourds, lettuce, beets, cabbage, and other produce from his garden (*Moretum* 71–9). Peasants almost certainly sold some of the wild plants they collected too (e.g., strawberries).[86] Cities with large populations of urban workers no doubt could generate a substantial market demand for vegetables. As with wine and oil, however, there would be gradations of quality. Pliny's discussion of vegetable prices (*HN* 19.54) indicates that even the urban poor bought vegetables but that some produce was well beyond their means. Once more it is difficult to even estimate the scale of urban demand or how appealing vegetables would have been compared with other foodstuffs. To what extent, furthermore, did small urban gardens compete with suburban and rural producers?

4.3.6 *Other produce*

Like flowers and vegetables, fruits and nuts were good choices as cash crops and potentially had an even greater 'range.' Cato the Elder's African figs, 'accidently' dropped in the senate, took only three days to reach Rome (Plut., *Cat. Mai.* 27). Varro mentions in passing the high prices fetched by fruit at the top of the Via Sacra at Rome (*Rust.* 1.2.10). Kron speculates that there was a "very large market for fruits and nuts" which went beyond the rich.[87] Neville Morley suggests that the demand for fruit (as well as vegetables and other "luxury foodstuffs") led to specialized food production in Rome's suburbium.[88] This is undoubtedly the case but it is important to remember, as discussed in Chapter 2, that fruit and nut trees required a greater investment of time and effort than vegetables and flowers.

There was urban demand for legumes as well. There is evidence for specialized lupin dealers (*CIL* IV 3423 and 3483) and legumes might be purchased as fodder as well as for human consumption. But, like grain, they were bulky and of relatively low value relative to their weight.

Firewood, charcoal, and lumber, finally, would have also been in high, steady demand as fuel, building material, and for the manufacture of a wide range of other goods. Firewood and lumber would be relatively difficult to transport but, for a farm near a city, Cato says it was especially useful to have an *arbustum* (plantation of trees) so one could sell firewood and twigs in town (*Agr.* 7.1). He advises the *paterfamilias*, if unable to sell firewood, to turn the wood into charcoal (*Agr.* 38.4). Varro praises the planting of elms in particular as especially profitable because they provided not only firewood but also support for vines, food for sheep and cattle, and building material for fences (*Rust.* 1.15). Props might also be sold to nearby farmers (*Rust.* 1.16.3). Columella, however, suggests that returns were fairly low. He indicates that one hundred sesterces per *iugerum* was a good return for woodlands (*Rust.* 3.3.3). As William V. Harris has convincingly argued, "by the second century BCE and even more in later times Roman estate-owners regarded many species of trees as useful assets, to be exploited and therefore to be propagated."[89] Robyn Veal's calculations suggest that the demand of Rome and other Italian cities for fuel in the form of wood and charcoal was tremendous.[90]

4.3.7 Linen

In some regions of Italy, where conditions were suitable, the cultivation of flax was a good choice for a farmer with the right kind of land. The Po valley, Campania (especially around Capua), and some coastal areas of Tuscany were known for their linen production.[91] Though not as common as wool, linen clothing was in widespread use and linen had several other important applications.[92] Unfortunately we know very little about the organization of the linen trade and it is consequently very difficult to estimate its scale. Columella warns against growing flax unless one is sure of a high yield and good prices (*Rust.* 2.10.17). It is also worth noting that the processing of flax into linen was a fairly labor-intensive process.[93]

4.4 The sale of the superfluous

As noted earlier, anything of value might be sold for money. Cato makes just this point in a notorious passage in which he recommends selling not only surplus grain (*frumentum quod supersit*), an old wagon (*plostrum vetus*), and old tools (*ferramenta vetera*), but also worn-out oxen (*boves vetulos*), blemished cattle (*armenta delicula*), blemished sheep (*oves deliculas*), and slaves who were old and sickly (*servum senem, servum morbosum*) (*Agr.* 2.7). His general strategy is to get rid of extraneous equipment as well as animals and slaves who consume supplies without providing any potential return. This differs from his approach to the sale of major commodities where he recommends ample storage in order to be able to wait for good prices (*Agr.* 3.2; see also Varro, *Rust.* 1.22.4 where storage capacity protects the farmer from being forced to sell).[94] As Plutarch's account of Cato's turn away from agriculture suggests, farmers probably tried to take advantage of non-agricultural opportunities if their land had suitable resources. In the *De agricultura,* for example, Cato mentions burning lime in partnership with a *calcarius* and later gives instructions for the construction of a limekiln and its use (*Agr.* 16 and 38). In Varro's first dialogue landowners are encouraged to exploit stone quarries and sandpits, if available, or to construct an inn for travelers if the location was suitable (*Rust.* 1.2.22–3). Even unused farm buildings might yield money if they were dismantled and the building materials sold. This was apparently common enough in the early Empire to prompt legislation to limit the practice.[95] One could also opt to be a rentier and lease some or all of one's agricultural land to shepherds or tenant farmers.[96]

4.5 Working for others

For many smallholders, occasional work on neighboring estates was probably a regular source of income. There seems to be general agreement both
that smallholders were underemployed and that larger estates required seasonal workers to supplement their core slave staffs at times of peak labor
demand. Erdkamp, for example, argues that:

> peasants and small farmers, who did not hire day-labourers, had to
> have at their disposal the human and animal labour required to work
> the land at peak times. This resulted in a significant seasonal underem-
> ployment of their labour in their own farm.[97]

The large villas, as Rathbone suggests, could exploit the "underemployment
of the neighbouring free peasantry" and so get away with carrying "no surplus labour."[98] Erdkamp speculates that "ceramics and other manufactured
goods were produced in the countryside rather than cities" because of the
available labor of the poor, but suggests that *agricultural* employment opportunities tended to come at those times of the year "when peasants had least
labour to spare."[99] However, since the harvest times of grain, grapes, and
olives are spread out over the summer and fall, a smallholder could easily
concentrate on grain cultivation, for example, and then work for wages during the grape and olive harvests. More persuasive is Erdkamp's suggestion
that hired labor was preferable to additional work on one's own property
because of decreasing marginal returns on the investment of labor. Pliny
the Elder's comment about the danger of really good farming (*HN* 18.38)
indicates that some (but not all) farmers made the sort of calculation that
Erdkamp envisions (i.e., that beyond a certain point, more work on one's
farm was not worth the effort). The agricultural writers tend to emphasize
the sheer amount of work to be done in agriculture as well as the dangers
of laziness and not doing enough. They rarely caution their readers against
devoting *too much* attention to their land. Very early in the *De agricultura,*
Cato discusses the scenario of a *paterfamilias* visiting his estate and finding
that not enough work has been done (*Agr.* 2). He lists typical excuses of the
vilicus (sick slaves, bad weather, etc.) and advises the *paterfamilias* to remind
him of the indoor tasks which can be accomplished when it is raining as well
as the rules governing what can and cannot be done on holidays. These ideas
recur in later agricultural literature. Virgil (*G.* 1.259ff.) emphasizes the tasks
which are permissible on holidays or suitable to periods of bad weather.
Columella (*Rust.* 11.2.91) notes that there is always some work that can be
done at night (such as sharpening tools or making handles for them). In a
discussion of viticulture, Columella suggests subdividing the vineyard into
relatively small plots in order to (among other reasons) make the amount
of necessary work seem less imposing (*Rust.* 4.18.2). In a later passage, he
stresses the importance of starting farm work at the crack of dawn and

not allowing laziness to slow down the workers (*Rust.* 11.1.14). Perhaps the agricultural writers considered this sort of advice necessary because they wanted the estate's slave staff, about whom they are mainly writing, to be exploited to its full potential (few slaves would have much incentive to work beyond the minimum amount they could safely get away with),[100] but depictions of agriculture in Roman literature rarely emphasize the potential for leisure (with the exception of shepherds and villa *owners*). One of the virtues of the ass, according to Columella, was that it was able to endure the neglect of an ignorant keeper (*Rust.* 7.1.2). In other words, the ass was especially useful because it required little time from its owner. Some crops were also praised for their ability to withstand neglect or relatively light labor demands (see Chapter 2). It may be best not to think of smallholders as underemployed so much as having considerable flexibility in deciding how to allocate their labor, how intensively to cultivate their land. Farm work was exhausting and there was nearly always something to do.[101] The appeal of hired work was no doubt the more immediate, tangible reward.[102]

The agricultural writers clearly expect farmers to have to (and be able to) hire outside workers. Cato assumes that the estate owner will have to hire *operarii*, *mercenarii*, and *politores* (*Agr.* 5.4; see also 4). Varro, in his discussion of agricultural labor (*Rust.* 1.17.2), says that farming is done by slaves, free men, or both and that, while very many poor people farm with their children, others employ workers (he singles out their use in the grape harvest and cutting of hay).

Men probably performed most of this wage labor with women more likely to be employed in tasks in and around the farmhouse. However, women in poorer rural families sometimes must have had to resort to work as day laborers.[103] Strabo indicates that women worked in the fields and even tells the story of a woman who, while working as a hired ditch digger with men and other women, gave birth and then went back to work for fear of losing her wages (3.4.17).

4.6 Moneylending

The sale of goods and labor were, of course, not the only ways for farmers to gain the cash they needed or desired. They could also borrow it (or earn money through making loans themselves should they have sufficient capital). However, as far as rural lending is concerned, direct evidence is incredibly sparse for Italy in this period. Cato the Elder warns the estate owner that his *vilicus* "should not lend seed, fodder, grain, wine, or oil and have no more than two or three households with which he lends and borrows" (*Agr.* 5.3). Certainly there are plenty of references to indebted farmers,[104] so there were sure to have been lively networks of rural credit but the creditors are rarely identified with any precision.[105] Patrons lent to clients, landlords lent (or extended credit) to tenants, and friends and relatives borrowed from one another.[106] In addition to formal loans, Cato's remarks indicate that there would also have been much informal lending and borrowing among neighbors. Creditors are likely to have been the wealthier rural households, artisans, and merchants. Comparative evidence certainly points in that direction. We should imagine loans both in kind and coin, the provision of credit for purchases, and, when the creditor was neither a relative or friend, higher interest rates.[107] Many of such loans were probably arranged rather informally.[108] There is no sign that Roman bankers had penetrated the countryside but, as it is clear many Roman farmers went bankrupt, many a wealthy farmer must have become wealthier thanks to loans.

4.7 Conclusion

As we have seen, farmers had many possible strategies to choose from in order to earn the money they needed for their own essential purchases. The wealthier estate owners, of course, had more and better options since they would have the time, appropriate land, and necessary capital to engage in businesses such as large-scale pisciculture which had complicated and expensive requirements. They could also simply decide to lease their land and leave the agricultural decisions (and some of the risk) to others. Poorer farmers probably relied heavily on seasonal day labor which the proliferation of slave-run estates would have made a readily available option from the second century BCE until at least the early Empire. With respect to the sale of plants, animals, and related products, many considerations and constraints would govern a farmer's choices. However, there is ample reason to believe that production and consumption of milk, cheese, and honey was substantial within Italy and that this consumption spanned both rich and poor. Our elite sources indicate some large-scale operations in terms of large or numerous flocks, giant cheeses, and rows of hives.[109] Clearly, however, honey production and to a slightly lesser extent milk and cheese production were well within reach of non-wealthy farmers. There would also have been considerable trade in firewood, wine, olive oil, flowers, legumes, fruit, nuts, poultry, and eggs. Though many smallholders no doubt regularly sold small quantities of surplus grain, there were better ways to earn income from agriculture.[110] Rome, and the other cities of Italy, enriched by a growing Mediterranean empire, increasingly demanded and could afford many rural products beyond staple cereals.

Notes

1 See, for example, Rosenstein (2008) for a discussion of the potential profitability of agriculture in the mid to late Republic.
2 See de Haas (2017) for discussion of geography and rural development especially with respect to accessibility of transportation infrastructure.
3 Tanning: Groot 2016, 59–60. Flohr (2013c) suggests tanning was "an entirely urban craft."
4 Sinews: van Driel-Murray 2008, 488.

5 Bone ash: Stern 2008, 525.
6 Fodder: Foxhall 2007, 82.
7 Profits of convertible husbandry: Kron 2000.
8 MacKinnon (2004, 225) notes the lack of "zooarchaeological data from peasant dwellings" and evidence for "small-scale exchange of animals or their goods" (247).
9 Meat consumption: MacKinnon 2004, 212. Corbier (1989, 224), drawing mainly on literary evidence, had already observed that the "use of meat seems to have been relatively banal and widespread, at least in the city."
10 Meat consumption: MacKinnon 2004, 215; Jongman 2007a, 191.
11 Increase in size of livestock: Riedel 1994 and MacKinnon 2004, 243.
12 Game: MacKinnon 2004, 212.
13 Evidence for the urban meat trade: MacKinnon 2004, 244.
14 MacKinnon (2004, 226) suggests "most of the meat consumed derived from public sacrifices." Kleijwegt notes (2002, 98) that it "is difficult to ignore (although extremely difficult to quantify) ... the profound effect sacrifice had on the scale of animal husbandry."
15 Forum Boarium: Holleran 2012, 95.
16 Fixed locations: Holleran 2012, 9.
17 Frayn (1993, 146) suggests the poor mostly ate "ham or salt pork," while MacKinnon (2004, 209) observes that there "is no mention of beef, veal, or pork consumption specific to the Roman rural poor," and speculates that the urban poor may have purchased sausages and "from minced-meat suppliers." See also Jongman 2007a, 193; and Kron 2002, 65.
18 Rural cattle: MacKinnon 2004, 215.
19 Meat yields: MacKinnon 2004, 225.
20 Livestock distribution: MacKinnon 2004, 248.
21 Competition: Jongman 2007b, 605.
22 Synergy between crops and livestock: Kron 2004, 124.
23 Military diet: Corbier 1989, 229.
24 Demand for hides: van Driel-Murray 2008, 490; Bishop and Coulston (2006, 247) suggested only forty-six thousand goatskins.
25 Pisciculture: Higginbotham 1997; Marzano 2013.
26 Regulation of meat consumption: in addition to the *lex Fannia,* the *lex Didia* (143 BCE), *lex Aufidia* (date uncertain), *lex Licinia* (before 103 BCE), *lex Aemilia Sumptuaria* (115 BCE), and *lex Cornelia* (81 BCE) all limited meat consumption directly or indirectly. See Zanda 2011, 120–6.
27 Jongman (2000, 188) suggests textiles would have been the third largest expense for most people after food and housing.
28 Sources of good wool: Jongman 2000, 189. See also: Kron 2002, 65.
29 Processing raw wool: Jongman 2000, 191 and 189. Fregellae was also a major textile producer prior to its destruction (Kleijwegt 2002, 92 n. 43). In the south, Tarentum, well known for its sheep, was also a wool-working center (Goodchild 2013, 206).
30 Spinning and weaving by women: Flohr 2013b, 76.
31 For further discussion, see Jongman 2000.
32 Maintenance and reuse of textiles: Flohr 2013b, 56.
33 Lactose intolerance: Sallares 1995, 94.
34 Sallares (2007, 34) notes that "nearly a fifth of the skeletons of the people who were killed trying to flee Herculaneum during the eruption of Mt. Vesuvius in AD 79 display morphological changes consistent with brucellosis, a disease generally contracted by consumption of infected animal products, particularly milk from goats."
35 Milk consumption: Wilkins and Hill 2006, 162; see also Curtis 2001, 399.
36 Milk and medicine: Wilkins and Hill 2006, 162.

37 For the medicinal uses of milk, see Plin., *HN* 28.125ff.
38 Spoilage: Thurmond 2006, 190.
39 Military diet: Garnsey 1999, 125.
40 For recipes involving cheese as an ingredient, see Cato, *Agr.* 76–85.
41 Diocletian's Price Edict suggests that cheese was inexpensive while Plutarch implies (*de Cup.* 2) that even poor men could afford cheese. Neither piece of evidence, obviously, pertains directly to Italy in the period being studied.
42 Transhumance: Thurmond 2006, 198.
43 Rennet: Thurmond 2006, 201.
44 Whey: Thurmond 2006, 205.
45 Yields: White 1970, 313.
46 Sheep and cheese: Thurmond 2006, 197. Sheep were easier to manage than goats, less expensive than cows, and offered arguably more uses. See also: MacKinnon 2004, 243.
47 Soft cheeses: Thurmond 2006, 207.
48 Keeping sheep: Garnsey 1999, 17.
49 Subsistence farmers and dairy products: Frayn 1979, 40.
50 Cheese market: Kron 2012, 172.
51 Butter and medicine: Curtis 2008, 385.
52 Depictions of hives and beekeeping: Crane 1999, 203; Larew 2002, 319.
53 Eumachia frieze: Larew 2002, 319.
54 Bees as control marks on Roman coins: for example, those of L. Roscius Fabatus in 64 BCE (Crawford #412/1).
55 Quintus Cominius Abascantus: D'Arms 2000.
56 Honey in cooking: Petron., *Sat.* 31 and 66; Cato, *Agr.* 76ff.; see also, on the recipes of Apicius, Solomon 1995, 120.
57 Thurmond (2006, 247) argues that "honey was an expensive commodity available only to the wealthy in the urban centres." See also Evans 1980, 142.
58 Honey used to preserve fruits and nuts: Mattingly 1996, 222; Solomon 1995, 129.
59 Perfumes and cosmetics: Reger 2005, 260; and Curtis 2001, 417. Plin., *HN* 21.76 (for women's skin).
60 Diocletian's Price Edict gives a maximum price of forty *denarii* for one Italian *sextarius* of best quality honey (a little more than half a liter). Second quality honey was capped at twenty-four *denarii*. By contrast, a pound of fresh cheese was capped at eight *denarii*. For comparison, a *modius* of wheat was capped at one hundred *denarii*, a day's labor on a farm at twenty-five *denarii*, and a *sextarius* of Falernian wine at thirty *denarii*. Again, the use of the Price Edict is problematic.
61 Knowledge of honey: Plin., *HN* 11.32ff.; Petron., *Sat.* 38.
62 Labor demands of apiculture: Thurmond 2006, 253.
63 Honey harvest: Reger 2005, 278; Columella, *Rust.* 9.14.1; Varro, *Rust.* 3.16.28.
64 On the sale of grain: Erdkamp 2005, 134.
65 Grain sales of small farmers: Erdkamp 2005, 135.
66 Urban sale of hay: Erdkamp 2005, 74.
67 Yields: compare with Varro (*Rust.* 1.44.1–2) who says Etruria yielded 15:1. Erdkamp, who provides an extensive discussion of the yield evidence (2005, 34–54), argues (38) that "[a]t best, Columella provides a trustworthy figure for poor soils; at worst, his estimate is not reliable at all."
68 On urban demand for grain, see Rosenstein 2008 and Launaro 2017. Heinrich (2017, 167) suggests that "for smallholders the small-scale production of garden vegetables, fruits for the table and animal products for nearby urban markets could have been more attractive than producing a cereal surplus ... Italian cereal production would be mainly focused on domestic consumption."

69 Wine from the neighborhood of Rome: Wilson 2009, 216. There are other difficulties. Purcell (1985, 17) suggests amphorae were not used for wine unless overseas transport was envisaged while Carandini (1989, 16) says lower quality wine would be stored in *dolia* rather than amphorae.
70 Wine consumption: Rosenstein 2008, 5.
71 Wine trade and its potential scale: Tchernia 1986; Launaro 2017.
72 High quality wine: Carandini 1989, 16.
73 Land requirements for wine and olive oil production: Jongman 2003a, 115.
74 Demand for wine: Rosenstein 2008. See Launaro (2017) for a critique of his argument.
75 Advanced sale of wine: Erdkamp 2005, 109; Plin., *Ep.* 8.2; Cato, *Agr.* 147.
76 Wine trade intermediaries: Purcell 1985, 11–13.
77 Grape syrups: Thurmond 2006, 248.
78 Labor demands of olive cultivation: Erdkamp 2005, 74. See also Mattingly 1996, 221.
79 Advanced sale of olives: Erdkamp (2005, 130) suggests it was probably "common practice" at least "among market-oriented landowners." See also: Cato, *Agr.* 146.
80 Annual olive oil consumption: Garnsey 1991, 84; Hitchner 2002, 72.
81 Rome's demand for olive oil: Hitchner 2002, 73.
82 The marketability of olive oil: Mattingly 1996, 225.
83 Trade in olive oil compared to trade in wine: Lafon 1993, 263.
84 Flowers and religion: Neudecker 2015.
85 Street traders and vegetables: Holleran 2012, 9.
86 Sale of wild plants: Witcher 2016, 471.
87 The market for fruit and nuts: Kron 2015, 168.
88 Demand for fruit and other foodstuffs at Rome: Morley 1996, 107.
89 Exploitation of trees: Harris 2013, 192.
90 Demand for fuel: Veal 2017.
91 Regions known for linen production: Gleba 2004, 34; Fredericksen 1959, 110.
92 On the use of linen, see Chapter 2 and Gleba 2004, 32. Pliny the Elder (*HN* 19.22–3) mentions dyed linen sails, flags, as well as the awnings of theaters, streets, and the forum.
93 Labor demands of flax processing and linen manufacturing: Aldrete et al. (2013, 149–51) estimate that it would require 675 hours of labor to produce the linen for a linothorax. Obviously typical linen clothing did not consist of ten layers of laminated fabric but even dividing the amount of time by ten (the number of layers in a linothorax of minimum thickness) yields 67.5 hours of work.
94 Forbes and Foxhall (1995, 75–6) observed a similar strategy at Methana where farmers held on to valuable goods that could be safely stored for long periods of time and only sold them when some major purchase was necessary.
95 Demolition: see Phillips (1973) on the *Senatus Consultum Hosidianum* (*ILS* 6043) of ca. 45 CE.
96 On tenancy: see Finley 1976; de Neeve 1984a; Kehoe 1989; Foxhall 1990; and Kehoe 1997.
97 Underemployed smallholders: Erdkamp 1999, 558. See also Evans 1980, 137; Jongman 2003b, 184; Rosenstein 2004, 65; and Launaro 2011, 182.
98 Surplus labor: Rathbone 1981, 15.
99 The rural labor supply: Erdkamp 2005, 84 and 82.
100 Some slaves would have had *peculia* however. See, for example, Varro, *Rust.* 1.2.17.
101 Kron (2017, 122) envisions the Roman countryside "teeming with small-holders ... often hard pressed to keep up with all of the demands of working their land well."
102 Rewards: work on the farm of another could offer illicit opportunities for personal gain through the theft of equipment or food. Columella (*Rust.* 8.10.3–4)

reports that some estates hired people to chew figs and then feed them to the thrushes they were raising. He advises against the practice because of the expense, partly caused by the workers eating the figs themselves.

103 Rural work opportunities for peasants and women: Erdkamp 2005, 82 and 88; Scheidel 1995.

104 Indebted farmers: Catiline's supporters (Cic., *Cat.* 8–10, 21), tenants (Plin., *Ep.* 3.19, 9.37). Varro (*Rust.* 1.17.2) mentions *obaerarii* (indebted workers) but associates them more with Asia, Egypt, and Illyricum. Columella (*Rust.* 1.7) seems to be referring to debts in a discussion of relations between master and tenants, going so far as to quote the *faenerator* Alfius.

105 The identity of rural creditors: Cicero, for example, describes many of Catiline's supporters as in debt but never says to whom they are in debt (*Cat.* 21).

106 Rural credit: Pliny the Younger and other landowners who leased out land became creditors when their tenants defaulted on cash rents (Plin., *Ep.* 3.19, 9.37). Horace (*Ep.* 1.7) tells the story of the wealthy patron Philippus, who lent money (in addition to a gift of money) to the auctioneer Volteius Mena so he could become a farmer.

107 Comparative evidence for rural credit networks: in a study of rural Chile, Charles Nisbet (1967, 73) found a wide variety of lenders but, because there was "little or no competition" among commercial lenders, very high interest rates. In Pakistan, Malik and Nazli (1999, 699) found that "[c]redit policy aimed at improving access of the small landowners and the poor ended up being diverted to the powerful large landowners" but it was difficult to determine actual interest rates.

108 On the nature of informal credit markets, see Nisbet 1969, 162.

109 Large-scale agriculture: Varro (*Rust.* 3.16.16) and Columella (*Rust.* 9.7.3) both say that three vertical rows of hives are enough.

110 On the market strategies of smallholders, see Heinrich 2017, 167 and Kron 2017, 115.

5 Farmers' markets, farmers' networks

5.1 Introduction

Having surveyed the external needs of farmers – what they could not produce themselves – and what they could sell or exchange in order to acquire those needs, it is now time to consider the issue of where and how these transactions took place. There are a number of different ways farmers could sell their surpluses and supply their external needs and they have different implications for the Roman economy. For example, prices can differ depending on whether smallholders tend to sell in urban markets or locally to traders.[1] Other factors include the medium of exchange for transactions and the timing of sales: before the harvest, at the harvest (when prices tended to be lower), or later (when the possibility of higher prices competed with the increasing likelihood of spoilage). It should go without saying that the evidence for rural exchange is poor since we rarely see beyond the interests of elite villa owners whose extant writings devote little time to economic issues, while the archaeological evidence sheds little light on the processes of exchange.[2] It should also be obvious that the modes of rural exchange in late Republican and early imperial Italy were complex and changing. The diversity of conditions for agriculture, discussed in Chapter 2, are only part of the reason for this. Cultural and political differences within Italy were profound despite a gradual trend towards integration.[3] And then there are the different economic strata within Roman society (from the owners of many large villas to landless, itinerant harvesters), their differing strategies, the non-agricultural professionals (artisans and traders) and *their* strategies, the places where these groups interacted and the times those interactions occurred, and, finally, the rules governing their exchanges. These are all complicating factors and, in many cases, very poorly understood ones at that. With these caveats noted, I want in this chapter to consider how farmers met their external needs through markets as well as other means (relatives, friends, neighbors, patrons, clients).

It may be useful to begin with a review of the actors involved in rural exchange. There was a broad spectrum of landowners. At the top were the extremely wealthy who owned many estates, scattered throughout the peninsula and increasingly beyond, which they cultivated by means of *vilici* and slave staffs (with additional hired labor) or leased to tenants (see below). These are the people we know best, both from extant literature and the remains of their villas. Sons or other dependents might also manage an estate or group of properties on behalf of a *paterfamilias*.[4] Even among the wealthy there was substantial variation, from the vast number of properties increasingly owned by the emperors to the relatively modest holdings of poorer senators and equestrians.[5] It is important to remember that sanctuaries and cities could also own land which they might lease in order to earn income.[6] And then there were the moderately wealthy, the local elite everywhere but Rome. These *decuriones* probably also owned multiple properties but within a more constricted area. They too would likely have had many slaves. Beneath this group would be a range of smallholders from the reasonably prosperous to those struggling at or below subsistence level.[7] Some might own a few slaves and command the labor of family members with which to farm a reasonably large piece of land (or scattered plots) while others would have had little land or additional labor. Over time, of course, an individual family's available labor would increase as children grew up and became able to contribute to household tasks, or decrease as members married, served in the military, became ill or disabled, or died. Knowledge and skills would be gained and lost. Similarly, a family's available land would vary with additions made through inheritance (gaining ownership or perhaps just a usufruct), marriage, purchase, or lease and losses due to the provision of dowries, sales, confiscations, and seizures (due, e.g., to defaulting on debt obligations). Survey archaeology indicates a wide range of rural settlement sizes throughout the period in Italy,[8] but it is hard to assign more than rough percentages to the different levels of personal wealth.[9] Some rural dwellers would own no land (or no agricultural land) even leaving aside those who were enslaved or under the authority of a *paterfamilias,* or in the employ of another farmer. These might include village-based artisans such as blacksmiths and potters as well as shepherds who made use of leased or public grazing lands. No doubt there were typically some landless poor surviving as tenant farmers or itinerant farm workers while other itinerant workers might appear seasonally from nearby urban areas. Passing through the countryside were also merchants, traders, itinerant religious professionals (the astrologers et al. the agricultural writers warn about), quacks, fugitive slaves, and thieves.

As to how goods moved among these individuals and groups, there are three modes of economic transfer: market exchange, reciprocity, and redistribution.[10] To understand the economic networks in which Roman farmers operated, it is necessary to discuss in turn transfers in each of these modes (though the distinctions between them can sometimes be blurry). That means considering markets and traders, but also farmers' relationships with family, friends, neighbors, patrons, clients, and government.

5.2 Markets

Peter Temin defines market exchange as comprised of transactions involving instrumental behavior and variable prices.[11] From the perspective of the Roman farmer seeking to buy or sell a physical object (as opposed to his or her labor), such transactions usually took place on a farm or in some sort of physical market (the weekly *nundinae*, permanent markets in towns, or annual fairs).[12]

Buying or selling on one's own farm had advantages and disadvantages. On the one hand, the farmer was saved the trouble of transporting the goods in question to or from a market, which would have cost time and money. There was also some risk in transporting goods any distance as it is clear that rural security was not a high priority of Roman officials.[13] Varro, admittedly writing at a particularly unstable moment in Roman history, says there were many excellent farms which it was not profitable to farm because of robberies (*Rust.* 1.16.2; see also 1.12.4). In addition, as Morley has pointed out, negotiating transactions on one's own estate allowed the owner to put the trader buying one's produce (or peddler selling his or her wares) at a disadvantage since the estate was the farmer's 'home turf' and this could make it awkward for the less socially respectable merchant to get the best possible deal.[14] Sale at the farm also enabled the farmer to avoid becoming too involved in what some viewed as a disreputable profession.[15] In certain cases, when a crop was sold before harvest, the farmer was also relieved of the responsibility for the gathering and processing of the crop.[16] That might be appealing since the farmer would not have to worry about organizing such tasks but there was always the possibility of outside workers causing damage to the farm and its equipment (Cato, *Agr.* 145.3 and 146.3 on damages caused by hired workers). Early payment might also be very appealing to those pressed by debts or in need of ready cash for other reasons.[17] On the other hand, a major disadvantage of transacting business on the farm would be the lower prices the farmer's goods might fetch compared to those of a physical market where there could be more competition among buyers or, at least, the seller would be more aware of prevailing prices and news affecting those prices. For example, a farmer selling 'at the farm gate' might not learn in time that the delayed arrival of the Alexandrian grain fleet or a damaging storm at Ostia had driven up food prices at Rome. By the same token, a farmer who bought from a traveling salesperson would not have the opportunity to compare wares and prices with other sellers.[18]

When buying or selling in markets as opposed to at home, the situation was somewhat reversed. The farmer looking to sell produce might have a chance to find a higher price thanks to competition among buyers and better knowledge of market conditions, but he or she undertook the risk and expense of transport. Wealth would obviously be a limiting factor here. Poorer farmers may not have had the time, beasts of burden, appropriate storage containers, or personnel to bring bulky commodities to a market. This may explain why we rarely hear of smallholders selling such goods (grain, wine, and olive oil) as opposed to vegetables.[19] Wealthier farmers more often would have had the luxury of choice. With larger surpluses (and more capacity to buy goods), merchants and traders may have been more likely to visit their estates. Such farmers could afford sufficient storage facilities to wait for good prices, as Cato and Varro recommend. They would also be more likely to have the manpower and beasts of burden to bring goods safely to markets. Wealthy farmers who maintained urban residences – which seems to have been the norm – could bring produce there to enjoy themselves or sell, if the right occasion arose.

In addition to formal sales between farmers and merchants, we should also envision a variety of rather informal sales. By this I mean a range of transactions from, for example, the purchase of a new dog from a passing shepherd (as Varro recommends, *Rust.* 2.9.5) to the sale of produce between neighboring farmers. These sorts of sales were probably quite common and, particularly when between neighbors, accomplished through in-kind exchange or credit.[20] With respect to more formal exchange, however, there seems to be a clear pattern. Wealthier farmers either sold to traders who visited their estates or transported their goods to markets themselves. They do not seem to have made much use of the *nundinae* for the sale of their produce.[21] Smallholders, by contrast, likely sold much of their surpluses at local *nundinae* and that is probably where merchants would have gone if they wanted to purchase them – it was hardly worth their effort to visit individual small farms (though the villages of smallholders might have been attractive stops).[22] However, as several scholars have pointed out, many sellers, selling a relatively uniform commodity (such as grain), would lead to the phenomenon known as 'perfect competition' which tends to keep prices low.[23] Smallholders would have little choice but to accept the prevailing market price for wine, oil, and grain (and no doubt legumes as well). As is (and was) widely recognized, farms in close proximity to substantial urban markets tended to have a considerable advantage over those further away. Unless more distant producers were able to ship their goods, those closer to town would have lower transportation costs and could sell directly to urban consumers. Unless a farmer was selling something like honey, which kept very well and had a relatively high value to weight ratio, it would be hard to exploit urban demand from a distance.

If we consider what farmers needed to purchase, the issue of transportation costs is less of a factor. Most of the goods a farmer might need to buy were cheaper and easier to move long distances. Merchants could bring salt, pitch, and iron tools to rural consumers at less expense than an equivalent value of, for example, grain could travel the same distance.[24] (Perhaps one of Cato the Elder's motivations for specifying places to purchase particular farm supplies (*Agr.* 135) was the knowledge that inferior items were readily available from itinerant peddlers whose wares would not stand close comparison with alternatives at *nundinae* or permanent markets.) There were exceptions, of course, like the mills Cato mentions, whose transportation costs could be substantial (nearly 75 percent of the purchase price in the case of the one from Pompeii, *Agr.* 22.3). Such large equipment was not essential, however, for individual smallholders.

Tenancy constitutes another kind of market transaction of particular importance in agriculture.[25] One must note, first of all, that tenancy can describe a wide range of practices and situations. While many smallholders and landless farmers must have rented land, more prosperous Romans often did as well. A five-year term seems to have been typical. The *locatio conductio* contract entailed a fixed cash payment and so implies that the renter would have to sell some of the produce to pay the rent.[26] Share cropping (in which the rent constituted an in-kind payment of a portion of the harvest) was also a possibility but could be less appealing to the landowner since it might require more supervision of the tenants at harvest time.[27] Varro's tale of Seius, who received five thousand pounds of honey each year as rent for his apiaries (*Rust.* 3.16.10) shows that a fixed in-kind payment was also possible. Columella (*Rust.* 1.7) mentions payments of money but also firewood and "other small additions" in a discussion of relations between an estate owner and tenants. What sorts of tenancy agreements were most common in any particular period or region of the Roman Empire (with the possible exception of Egypt)[28] remains uncertain but it now seems clear that at least some form of tenancy was always an option in Roman Italy.[29]

Tenancy was a way to manage risk, capital, and labor. For the landlord, renting land to a tenant could help ensure a steady stream of revenue from a property for which perhaps he or she did not have sufficient manpower to cultivate him- or herself. Except in cases of *force majeure* (e.g., severe drought or warfare), the landlord was entitled to the cash rent even if the tenants proved to be indifferent farmers.[30] If the tenants failed to pay the rent (or were unable), the landlord could seize and sell their tools, livestock, and slaves. Of course, as Pliny the Younger observed (*Ep.* 3.19), doing so might impair the ability of the tenants to properly cultivate the land, leading to further difficulties in making rent payments or damage to the land. Landlords were obligated to provide the requisite infrastructure for the tenants to use (e.g., storage containers) and might have to lower rents under some circumstances.[31] It was obviously highly advantageous for a landlord to have reliable, long-term tenants (and vice versa)[32] and such considerations would lead their relationships to have a broader, non-commercial dimension.[33]

5.3 Reciprocity

Reciprocal exchange involves the giving and receiving of goods and services according to "social obligations and traditions."[34] Among the Romans the conventions of patronage and friendship shaped these kinds of exchange. This is the sphere of gifts and favors and, though we rarely hear of it, Roman farmers of all varieties must have been as heavily engaged in them as the urban elite we know so much better. Rural Romans met for many reasons other than to conduct commercial transactions. Even the *nundinae* were not exclusively for the purpose of market exchange. I have already mentioned the numerous rural religious festivals that were celebrated. Many people traveled to nearby (or even fairly distant) towns to witness a spectacle (gladiatorial shows or chariot races), engage in politics or be party to legal proceedings,[35] or they visited the estates or houses of relatives, friends, patrons, or clients for the sake of leisure, advice, or their health. These visits helped establish or maintain ties of friendship, patronage, and marriage among Romans of varying wealth, status, and profession. These relationships in turn offered farmers other avenues by which to acquire what they needed (or desired).

Judging from the agricultural writers, neighbors were especially important for farmers. As discussed earlier, Cato urged the farmer to be good to his neighbors in order to more easily sell his produce, contract out work, and hire laborers (*Agr.* 4). Essentially, the farmer needed to establish relationships based on trust with neighbors in order to facilitate commercial transactions,[36] but Cato also notes that, if one's neighbors liked you, they would help you build with their labor, beasts of burden, and even timber, by implication for free. Friendship facilitated reciprocity as well as trade. Varro, perhaps reflecting the many disruptions in rural Italy in the last decades of the Republic, is more suspicious of neighbors. Though he does mention the possibility of buying and selling from them (*Rust.* 1.16), he emphasizes the importance of clearly marked boundaries to prevent quarrels between *familiae* and lawsuits over property rights. The character of Scrofa recommends planting pines, cypresses, or elms (*Rust.* 1.15). Unfortunately, even a neighbor's trees could cause trouble. Varro notes that oak or walnut trees near one's land can have a negative impact on certain crops (*Rust.* 1.16.6). This ambivalence is not unusual. Horace, himself an estate owner, also makes a passing reference to neighborhood quarrels in a discussion of farms (*Ep.* 2.171). Pliny the Elder, in one of his polemics against greed, refers to farmers stealing land from their neighbors (*HN* 2.175), but he later calls Cato's advice about cultivating the goodwill of neighbors self-evident (*HN* 18.44). Columella subscribes to Cato's advice as well, noting that many had abandoned their homes due to the injuries inflicted by neighbors and complaining about the problems caused by an (unnamed) neighbor of his own (*Rust.* 1.pr.5–7). The settlement of Romans on confiscated lands throughout the Italian peninsula certainly increased tensions between the remaining locals and their new neighbors.[37]

The wealthiest Romans had friends and relatives scattered throughout Italy and beyond. Smallholders, however, would have been much more geographically confined in their society. Their friends would be neighbors and, as land was divided by wills or given as dowries, they would be close to relatives too. (However, as noted in Chapter 2, it is entirely possible that extended rural households were quite common.) Smallholders living in villages would also have friends and relatives nearby who might be willing to help them with loans, gifts, partnerships, and so forth.[38] Of course, as any Roman would know, even a twin brother could turn into a dangerous enemy. Still, Luuk de Ligt is undoubtedly right to have argued that peasants "could generally count on receiving assistance from kinsmen, friends and neighbors."[39]

Patrons could also provide assistance to smallholders and other farmers, particularly with loans and legal help. Erdkamp has argued that wealthy estate owners tended not to lend money to the "rural masses" because "loans to the rural poor were not a very attractive investment."[40] I would argue, however, that such loans, while not reliably lucrative from a financial perspective, may still have had great appeal to (some) wealthy farmers. The (perhaps likely) default of the peasant might allow his or her creditor to take possession of land, livestock, and equipment or exert greater control over his or her indebted client.[41] Despite the abolishment of debt slavery (*nexum*), creditors could still wield tremendous power over insolvent debtors. Some late Republican politicians were able to raise armies out of their tenants and debts were likely a factor that helped compel them to fight on behalf of their creditors.[42] Commercial relationships could facilitate reciprocal ones. Pliny the Younger demonstrates this in his treatment of the merchants who had purchased his grape harvest in advance. Since the harvest had apparently not gone nearly as well as anticipated, Pliny gave the merchants all rebates of a portion of what they had paid. But, as he explains to Calvisius Rufus (*Ep.* 8.2), he gave larger rebates to those who had made larger purchases and those who had paid what they owed in a timely fashion. In gratitude for the size of their transactions with him or their conscientiousness in paying debts, Pliny paid some of the merchants more than he had to in the hope that they would continue to do business with him. (Pliny also notes that the gesture had garnered him much praise in the community.)

Reciprocity and commerce can also go hand in hand. Horace's Philippus gives his auctioneer client a gift of money *and* a loan (*Ep.* 1.7.80–1). Some estate owners had as clients their freed rural slaves. While it is unlikely that many agricultural slaves were manumitted, slave *vilici* sometimes had *peculia* with which they could have purchased their freedom.[43] Freed slaves, of course, became the clients of their former owners.

5.4 Redistribution

Redistribution usually refers to transfers of resources accomplished by a government through taxation.[44] The *frumentationes* at Rome, which gave provincial tax grain to eligible Romans, is a good example of this kind of transfer. Roman farmers stopped paying *tributum* (a kind of property tax) after 167 BCE (*HN* 33.56 and Plut., *Aem.* 38) though they might have to pay other taxes.[45] Other forms of redistribution, however, factor heavily into late Republican history: confiscations of land (most notoriously during the proscriptions but also in a sense by the Gracchan land commission) and land distributions as part of various settlement schemes (colonies and viritane distributions). But the biggest and most consistent redistributive process in the period under consideration was the wealth pouring into Italy from an expanding empire in the form of tax revenue and spoils. Its impact on Roman agriculture was partially indirect (in that the Roman authorities were not always using it deliberately to aid farmers) but nevertheless this wealth had a tremendous effect. The Romans exploited the influx of slaves on their villas, exploited Hellenistic and Punic agricultural knowledge, and imported new crops and animals too. Veterans invested some of the wealth they gained from conquest in the farms on which they were settled and new roads could bring those farms closer to markets that were also flush with imperial money.[46] Aqueducts, not exclusively for urban baths and fountains, also contributed to the new agricultural infrastructure while improved harbor facilities and (eventually) safer seas facilitated Roman exports. Farmers in Roman Italy were sometimes the victims but more usually the beneficiaries of the Empire's redistributive processes. In the final chapter, I will consider the interaction between farmers and broader economic and political developments in more detail.

Notes

1 Selling locally versus in towns: de Ligt 1993, 213–14.
2 Archaeological evidence: MacMullen (1970, 333) notes that rural periodic markets are nearly invisible archaeologically.
3 Diversity of Italy: Horsfall 2001.
4 In his *Pro Sexto Roscio Amerino,* Cicero defends a man who had managed thirteen of his father's farms.
5 Variations in wealth: Scheidel and Friesen 2009, 75–84.
6 The land of sanctuaries and corporate bodies: Green 2007, 20 (on the sanctuary of Diana at Aricia); Cic., *Fam.* 13.76.2 (for the sale of land *in agro Fregellano* apparently owned by a municipality). On *ager publicus* owned by communities, see Roselaar 2010, 136ff.
7 Smallholders at subsistence level were not on the verge of starvation. As Clark (2007, 22–3) helpfully points out, "subsistence income" is "the income that just allows the population to reproduce itself" and "in almost all Malthusian economies the subsistence income considerably exceeded the income required to allow the population to feed itself from day to day."
8 Rural settlement sizes: Launaro (2011, 157) suggests, based on the analysis of survey data, that "'villas' and 'farms' ... tended to develop along similar trajectories, rather than being mutually exclusive." Both kinds of settlement were common, and became more common in most parts of Italy, from the late Republic into the early Empire. Villages, however, "remain a highly neglected category of settlement" (Witcher 2016, 475).
9 Assigning percentages to different categories of wealth: Scheidel and Friesen 2009; Longenecker 2010, 36–59.
10 Modes: Temin 2001.
11 Market exchange: Temin 2001, 172.
12 Markets and fairs: MacMullen 1970; Gabba and Coarelli 1975; de Ligt 1991 and 1993; Frayn 1993; Lo Cascio 2000; and Ker 2010.

13 Rural security: Lintott 1986, 129. Suetonius reports that at the beginning of Augustus' reign free and enslaved travelers were being abducted and forced to work on estates (*Aug.* 32).

14 Status of owners and merchants: Morley 2000, 218–19.

15 Disreputable professions: Cicero (*Off.* 1.150–1) lists fishmongers, butchers, fishermen, hirelings, and moneylenders as practicing vulgar professions. Traders were problematic as well. See also Bond 2016.

16 Passing on risk and transport costs: Morley 2000, 217–18.

17 Plutarch (*On Borrowing* 832) suggests it was those who were in debt who would be forced to sell their crops prior to harvest, by implication at some loss.

18 It does not seem implausible to speculate that some wealthy estate owners, who dominated a particular rural area, might 'encourage' local smallholders to buy solely from them or limit the access of traders of whom they did not approve.

19 Smallholders selling grain et al.: Erdkamp 2005, 134–5.

20 Commercial exchange with neighbors: Cato (*Agr.* 2.5) mentions accounts of money, grain, wine, and oil. He also refers to communal milling (*Agr.* 136) with payments in kind.

21 Wealthy farmers and the *nundinae*: Morley 2000, 216.

22 Smallholders and the *nundinae*: Morley 2000, 216–20.

23 Perfect competition: see discussion in Hollander 2015, 164–6.

24 Traveling salesmen: the *Digest* (14.3.5.4) discusses *circitores* who went around selling clothes and slaves sent by a shopkeeper to purchase goods. Cicero (*Clu.* 40) mentions a traveling *pharmacopola* (drug seller) who visited many *fora*.

25 Tenancy: see, for example, Finley 1976; Foxhall 1990; Kehoe 1997; Johnston 1999, 62–8; Kehoe 2007b and 2012.

26 Cash rents: Kehoe (1997, 148) notes that, at least in the legal sources (which admittedly focus on the problems of the Roman elite), "cash is almost universally assumed as the form of the rent payment."

27 Share cropping: Pliny the Younger (*Ep.* 9.37) discusses the tradeoffs involved in a switch to share cropping.

28 Leases in Egypt: Martin et al. 2014.

29 On the history of Roman tenancy, see de Neeve 1984a and de Ligt 2000.

30 On the remission of rent, see, for example, Frier 1989–1990; Kehoe 2012, 118; and *Dig.* 19.2.15.

31 Reducing rents: Pliny (*Ep.* 9.37) mentions having lowered rents during a lease term. Ulpian (*Dig.* 19.2.19.2) discusses the landlord's obligations with respect to equipment.

32 Long-term tenants: Columella (*Rust.* 1.7.3–4) suggests that conventional upper-class wisdom considered very long-term tenants (i.e., present from birth) extremely desirable.

33 Foxhall (1990, 111) notes that, among other things, "tenancy offers tenants access to ... power networks."

34 Reciprocity: Temin 2001, 170. Blanton (2017, 293) proposes limiting the definition of reciprocity to "extra-mercantile exchanges in which principles of return, gratitude, and counter-transmission are invoked."

35 Travel: for example, Tacitus (*Ann.* 4.62–3) mentions that many of those killed or injured in the collapse of an amphitheater at Fidenae in 27 CE had come from Rome. In 59 CE a fight broke out between Pompeians and visiting Nucerians at a gladiatorial show. Many wounded Nucerians subsequently made the much longer trip to Rome, seeking redress from the emperor (Tac., *Ann.* 14.17).

36 *Amicitia:* Verboven (2002, 342) emphasizes the importance of *amicitia* for the creation of the trust needed for credit transactions.

37 Tensions: Keppie 1983, 101–32.

38 On rural partnerships, see Lirb 1993.

39 Help for peasants: de Ligt 1993, 131. Kron (2017, 136) discusses the reasons larger landowners needed "to integrate themselves into a broader rural community of small farmers."
40 Wealthy estate owners' reluctance to lend money to peasants: Erdkamp 1999, 558.
41 Lending to peasants: Columella (*Rust.* 1.7) and Pliny the Younger (*Ep.* 3.19 and 9.37) both refer to the debts of tenant farmers though those debts may be due to arrears of rent rather than loans *per se*. Cato (*Agr.* 149) mentions the possibility of debts created by property damage in a lease of pasture.
42 Armies of tenants: Hollander 2007, 71–5.
43 Rural manumission: see Treggiari (1969, 106) for doubts about this practice. Varro twice refers to *peculia* for agricultural slaves (*Rust.* 1.2.17 and 1.17.5).
44 Redistribution: Temin 2001, 170. There can be other agents of redistribution. For example, within the early Christian communities of the eastern Mediterranean (Blanton 2017, 290–3).
45 Taxation: the *tributum* was temporarily reinstated during the civil wars of the late Republic (Plut., *Aem.* 38 and Cass. Dio 46.31). There were several other taxes which continued to be collected throughout the period including customs dues, rents on public lands, and *scriptura* (fees for the use of public pastures).
46 Roads and markets: however, as Witcher observes (2017, 41), "investment in roads and rural infrastructure did not automatically or instantly integrate rural communities or stimulate economic development."

6 Farmers in Roman economic history

6.1 Introduction

Many Romans took pride in what they grew themselves, just as people do today, and occasionally they mocked those who bought things they could have produced for themselves. As the preceding chapters demonstrate, this does not mean they were actually pursuing self-sufficiency. Some certainly advocated in favor of the efficient use of available resources, which could lead a farmer to be self-sufficient in a particular good or goods, and there were compelling reasons to engage in polyculture, but a host of factors led Roman farmers to rely, to varying degrees, on markets, reciprocal relationships, and (in some cases) the redistributive power of government to meet their needs. In light of this fact, it makes far more sense, when discussing the economic behavior of Roman farmers, to think in terms of degrees of dependency. The use of the word 'self-sufficient,' even in a limited sense, leads us to ignore or underestimate the relationship between Roman farmers and the economy.[1] In this final chapter I want to consider this relationship by first asking what levels of dependency various kinds of farmers had and then what the implications of that dependency might be. If smallholders were highly dependent on the market, then changes in the market could have a serious effect on them. Dependency is best considered in three dimensions in that there is dependence on the market but also, as discussed in Chapter 5, dependence on reciprocal relationships and redistributive channels. While ample evidence shows the Roman elite engaging in all sorts of reciprocal relationships, we can only guess that it was also an important facet of the economic lives of poorer Romans and Italians.[2] With respect to redistribution, prior to the *alimenta* there is little evidence for sustained efforts in support of agriculture.[3] Roman farmers tended not to be direct victims or beneficiaries of redistributive wealth transfers except at the beginning or end of their agricultural careers (i.e., when receiving land as part of a colony or viritane settlement or when their lands were confiscated).[4] Thus, in the following section, I will concentrate on market dependency.

6.2 Degrees of market dependency

The degree to which any particular Roman farmer was dependent on the market would, of course, vary over time as his or her circumstances changed. But we can still plausibly speculate about relative degrees of dependency for different groups of farmers. Let us consider the general situation faced by four rough categories of free farmers: the elite (i.e., very wealthy landowners), the moderately wealthy, smallholders, and landless agricultural workers.

6.2.1 Elite farmers

The wealthiest Romans are the ones we know best and they had the potential to be the least dependent on the market. This is because such individuals could own large amounts of land, providing them with the ability to grow a wide variety of crops, raise large quantities of livestock, have access to sufficient amounts of ancillary resources (woods for fuel and pitch production, salt pans, etc.), and command the labor of enough slaves to actually perform all the necessary work to operate an array of widely scattered estates and transport goods between them. However, there is little indication that wealthy farmers behaved this way. To begin with, they appear to have relied on hired workers to supplement the labor of their slave staffs and in some cases tenants to work their lands. Thus they depended on the labor market. Wealthy estate owners, furthermore, seem to have been eager consumers of an array of goods not produced within their households, including expensive imported wines, textiles, building materials, and furnishings. Even some of their slaves were purchased. Leaving aside the vast imperial *patrimonium,* only one (fictional) individual seems to have had the resources for complete agricultural independence: Trimalchio.[5] The agricultural writers, all wealthy men, do not advocate an extreme independence from the market but instead acknowledge its importance for sales and purchases. Consider, for example, Pliny the Younger's letter to Gallus (*Ep.* 2.17), in which he describes his seaside Laurentian villa. While he mentions many aspects of its productivity including wine (*lata apotheca*), grain (a *horreum*), multiple gardens, access to firewood, seafood, milk, and livestock, he also notes that one can get supplies from Ostia or a nearby village. Even though Pliny's villa has its own baths, he mentions the three nearby bathing complexes that could be used instead. Nevertheless, the *potential* for low levels of dependency is certainly important. Wealthy farmers, provided they were not servicing substantial debts or actively involved in politics (always a cash-intensive enterprise), could endure fluctuations in market prices because, while they *did* generally consume purchased goods, fewer such purchases were *essential*. They would also often be in the position to exploit fluctuations in the market. Finally, it must be emphasized that, although many elites profited from agriculture, there were strong non-economic incentives for them to invest in land and potentially much more lucrative non-agricultural moneymaking strategies available.[6]

6.2.2 *Moderately wealthy farmers*

Farmers who were well above subsistence level but lacked resources on the scale required for senatorial or equestrian status constitute our second category. These individuals would be more dependent on the market and for more essential items. Though they probably owned multiple plots of land and quite a few slaves, they would still need hired workers (or tenants) to cultivate those properties efficiently and they might not have the full range of necessary kinds of land to supply all their essential needs in sufficient quantities. It might not take much land to grow sufficient amounts of wine and olive oil for a large *familia* but many would lack the ability to produce their own iron tools, clothes, ceramics, salt, etc. Furthermore, while textual evidence sheds little light on this class of farmer, if the excavations of rural sites across Italy are any indication of the behavior of such people, they did not behave in a manner consistent with the pursuit of low market dependency. The ceramic finds from villas suggests they were purchasing wine, fish sauce, lamps, and perfume, while iron tools, glass, and other goods also regularly turn up. They had the *potential* to reduce some expenditures by relying on domestic production but the distribution of fineware and amphorae suggest they did not choose that approach. Many of them would still have had to purchase expensive equipment and other items essential to the cultivation, processing, and preservation of food.

6.2.3 *Smallholders*

The vast majority of Roman farmers must have been relatively poor and they were much more dependent on markets than wealthier farmers simply by virtue of having much less access to land and labor. While wealthier farmers might make up for the deficiencies of one plot of land with the produce of a better or differently situated piece of land elsewhere, peasant farmers were far less likely to have a diversified portfolio of land types. They would need to purchase tools, ceramics, ancillary supplies such as pitch and salt, as well as, in most cases, some of their clothing. Those whose land was close to Rome, another major city, or a port would have a considerable incentive to exploit urban demand for flowers, vegetables, milk, cheese, or honey. Such products could be highly lucrative and some could be produced on relatively small plots of land. But, of course, those sorts of specialized production relied on the market even more and smallholders would face competition not only from other smallholders but also from wealthier farmers who were, as our sources make clear, very interested in pursuing these kinds of opportunities. In this class would have been many who lacked ownership of enough land to support their basic food needs and such farmers might have to lease land, putting themselves at the mercy of the rental market. As discussed in Chapter 4, another tempting strategy would be occasional day labor. Just as smallholders would rely more on market relationships than the wealthy had to, they would also be more dependent on reciprocal relationships both with friends and relatives at their own socio-economic level, who might share food, equipment, or their labor with them, and patrons who could help them with legal problems, loans of money and seed, or even access to seasonal employment. The key point is that Roman smallholders were more vulnerable to market forces than is usually appreciated even when they grew most of the food they ate.

6.2.4 Landless farmers

Those who work the land are farmers even if they own no land.[7] Within this category were the landless rural poor who got by engaging in day labor as well as tenants who paid rent to work land owned by others. Some may have worked as shepherds, tending to the flocks of others, their own, or a combination of the two on public or private lands.[8] Such farmers were the most dependent on markets (and reciprocity) since nearly everything they would consume came to them by means other than domestic production. Also within this category would be predominantly urban day laborers who went into the countryside occasionally to exploit employment opportunities during periods of peak labor demand (i.e., harvests). These people would also have been very dependent on the market and redistributive channels in some instances, especially those residing in Rome who might benefit from the *frumentationes*, other handouts, or subsidized prices for some essential commodities.[9] Such urban workers could be a threat to those rural farmers who relied on part-time work on the land of others to supplement their income. A downturn in urban building might lead to increased competition for rural workers and depressed wages. An increase in the rural population, by increasing the labor supply, would, *ceteris paribus*, depress wages. Regardless of where they lived, those who owned no land of their own were the most vulnerable group of farmers, at the mercy of markets and employers, and likely having quite limited access to credit.

To sum up, those most capable of extremely low levels of market dependency did not seek such independence (although the *potential* for independence had an obvious appeal) while all other categories of farmer, for the reasons laid out primarily in Chapter 3, must have relied on markets, albeit in different ways and to different degrees. It follows from this that the agricultural economy and the broader Roman economy were more heavily interrelated than is generally appreciated since non-wealthy farmers constituted the overwhelming majority of the Roman population.

6.3 Farmers in the Roman economy

If Roman smallholders, instead of being self-sufficient, independent cultivators, were in fact quite reliant on the market, what, if anything, does that mean for our understanding of the Roman economy? To conclude, I would like to consider the interplay of Roman economic developments and the behavior of farmers in Italy during the late Republic and early Empire. Obviously, changes in population and urbanization can have a huge impact on agriculture since they affect the labor supply and demand for produce, two critical variables for farmers who rely on the market to any significant degree. This makes the ongoing debate over Roman demography all the more frustrating. The low, middle, and high counts entail not just different population numbers but differences in *when* (or if) the population grew or shrunk. Similarly, how, when, and where agricultural practices changed, issues still rather uncertain in some regions, have major implications for Roman economic history. A more densely populated and intensively cultivated countryside entailed a much greater demand for all sorts of agricultural tools, supplies, and expertise. The considerable regional variation within Italy makes generalizations based on limited evidence dangerous and some areas have received much more attention than others.[10] These problems make any agricultural history of the late Republic and early Empire regrettably tentative at this point. As our picture of Roman agricultural practices comes into sharper focus with more excavations, surveys, and modeling,[11] it will be important to consider not just such factors as urban demand for food and fuel, access to land, and the productive potential of the countryside but also the aggregate demand of farmers for the means of cultivation, processing, and storage. To that end, I offer the following sketch of developments across three periods:

Period 1: the second century BCE;
Period 2: the first century BCE prior to the reign of Augustus;
Period 3: the early Empire from Augustus to the late second century CE.

6.3.1 The second century BCE

The second century BCE easily poses the most challenges for any attempt to understand the development of the Roman agricultural economy, with most debate focusing on understanding the motivation behind Tiberius Gracchus' agrarian legislation and subsequent attempts at reform. Rome recovered fairly quickly from the devastations of the Second Punic War (i.e., military mortality and the destruction of property in Italy). Thanks to the Carthaginian war indemnity and other new sources of revenue, the Roman government soon had ample resources and was able to undertake all sorts of expensive projects in Rome and its environs. After winning the Third Macedonian War (167 BCE), the Romans were able to forgo the direct taxation of their citizens' land.[12] That same year, the sack of Epirus brought about the enslavement of one hundred and fifty thousand people. It is likely that the size of the Roman slave population grew substantially during this period.[13] The management of this population proved challenging. In addition to the two major slave revolts in Sicily, there were problems on the Italian mainland as well, at Setia in 198 BCE (Livy 32.26), in Etruria in 196 BCE (Livy 33.36), at Minturnae in 143 BCE (Livy, *Oxy. Per.* 53), and elsewhere later on (Diod. Sic. 36.2). In the early decades Rome settled many citizens in new colonies (there were also viritane distributions), mainly in the north and south.[14] Thanks to confiscations from Hannibal's Italian allies, Rome had plenty of land to give out.[15] Roman veterans were able to start many farms and, sometimes greatly enriched by plunder, could afford the livestock and equipment to run them well. At Rome, there was a growth in private wealth that, combined with more intense social and political competition, led to increased demand for expensive foodstuffs. Though some of the food and drink that was in high demand came from overseas, other things could be produced locally. Thus *pastio villatica* began to grow in importance. A series of sumptuary laws soon followed.[16] While we may doubt their overall effectiveness,[17] in the short term these measures probably hurt the profits of those who had decided to specialize in the production of such foods by (at least temporarily) suppressing demand. At the same time, the increased population of Rome drove up demand for locally grown vegetables, fruits, and other (non-luxury) foodstuffs. Farms near Rome now could be more lucrative and, as the government had not yet started to consistently regulate the grain supply,[18] even grain might sometimes bring a good profit. Land within easy transport distance of Rome became more expensive. Road-building (or the maintenance of existing roads) helped integrate markets and would tend to increase the value of land in the vicinity.[19] At Rome officials built new markets and other facilities conducive to trade (e.g., Livy 35.10.12 and 40.51). There are also signs of increased monetization in the countryside.[20] Coinage can facilitate exchange more easily than most commodities but loans calculated in coin instead of kind could increase the difficulty of repayment for farmers.[21] This seems also to have been a period of increased interest in Hellenistic and Punic agricultural practices, leading to more efficient production.[22] There was likely much competition among farmers, not only over profits but also over the appearance of their fields, orchards, tools, and facilities; hence Cato's warnings about expenses. Overall demand for agricultural equipment grew not just because of rural population growth but also because some farmers intensified cultivation.

The first half of the second century was almost certainly a time of growth in the Roman population and its standard of living, especially after the 190s when, in addition to the slave uprisings, there seems to have been a financial crisis (in 193 BCE) leading to new legislation imposing Roman lending regulations on the Italian allies (Livy 35.7). In the latter half of the century, however, problems quickly emerged. The boom in the free population of Italy led,[23] thanks to partible inheritance,[24] to an increase in rural poverty.[25] The increased number of slave-run estates in parts of central Italy probably contributed to the problem.[26] Rosenstein argues that in this period military service also impoverished smallholders by removing labor from farms.[27] This led to the crises of the late second century as poor Romans supported the land redistribution proposals of Tiberius and Gaius Gracchus and, increasingly falling below the minimum property qualification for military

service,[28] found a new ally in Gaius Marius when he abandoned those qualifications in order to raise an army to fight Jugurtha (Plut., *Mar.* 9).[29] As
Rosenstein persuasively argues, the rural supporters of the Gracchi were not
men who had "been forced from their farms" by conscription and slave-run
estates but were "simply the inevitable result of too many people attempting
to start out in life with too little land."[30] A decline in the number of building
projects in this period would have made matters worse for poorer Romans
looking to supplement their agricultural income.[31] Despite the violent response of conservative politicians to the legislative agenda of the Gracchi,
their land commission did manage to settle many Romans on new farms.[32]
A series of subsequent agrarian laws soon fully privatized the *ager publicus*
that had been distributed.[33] The institution of subsidized grain at Rome,
meanwhile, changed the incentives of farmers near Rome, making cereals
less appealing as crops. At the same time, wine production for export, especially to Gaul, took off and lasted into the early Empire.[34] Campania and
Etruria were at the forefront of this trend.

6.3.2 The first century BCE prior to the reign of Augustus

Developments are clearer after 100 BCE if only because they become consistently chaotic. Wars raged across the peninsula in this period, most notably
during the 80s (which featured the latter part of the Social War and fighting between the supporters of Marius and Sulla) and the 40s (intermittently
from Caesar's invasion in 49 to the Perusine War in 40 BCE). These wars and
other conflicts (especially the revolt of Spartacus) damaged farm property
and disrupted cultivation. A second major problem was the insecurity of
land tenure thanks to the proscriptions of the 80s and 40s and other, less
substantial acts of confiscation and redistribution which were carried out or
merely proposed. This, along with damage cause by warfare, led to fluctuations in real estate prices and probably inhibited investment in land (both
in the sense of depressing demand for real estate and incentives to invest in
farm buildings and processing equipment). These factors, and other events
occurring outside of Italy (see, e.g., Cic., *Imp.* 19, for the damage caused by
Mithradates' invasion of the province of Asia Minor), played havoc with the
financial market. There are many signs of debt crisis. C. Julius Caesar Strabo
attempted to provide some relief, perhaps in 90 BCE (Cic., *Att.* 5.21.13). In 89
BCE A. Sempronius was killed by moneylenders after enforcing an old law on
interest rates (App., *B.Civ.* 1.4). There were further attempts in 88 and 86 BCE
(Plut., *Sulla* 8 and Vel. Pat. 2.23.2). According to Cicero, debtors made up
the bulk of Catiline's supporters in 63 BCE. There were more problems in the
late 50s and 40s. The senate capped interest rates (again) in 51 BCE (Cic., *Att.*
5.21.13) and Caesar provided debt relief in 49 and 48 BCE (Dio 41.37–8 and
Plut., *Caes.* 37.2). Nevertheless, wealth continued to pour into Rome and
urban demand remained strong. With the continuation of subsidized grain
and eventual shift to its free distribution to tens of thousands of eligible

residents,[35] there were fewer incentives to grow grain near Rome.[36] This is not to say that grain cultivation in the area ended altogether – imports could not provide all that the city needed[37] – but grain became less important even though occasional disruptions in supply might lead to healthy profits.[38] *Pastio villatica* was becoming more important, as Varro's writings, additional sumptuary legislation, and the remains of villas demonstrate.

Other developments in this period probably include the growing use of tenants which reduced the risks to landowners, and assured cash revenue.[39] Among the elite, demand for coinage increased due to its portability, a very appealing feature in a time of great political violence and corruption.[40] In 49/8 BCE Caesar limited the amount of cash anyone was allowed to hold to sixty thousand sesterces (Dio 41.37–8) and also apparently required investment in Italian real estate (Tac., *Ann.* 6.16).

Aside from *pastio villatica,* livestock must have been a very appealing investment for farmers in this period. Their mobility rendered cattle safer than crops and pastures were less susceptible to damage in warfare than farm buildings. Those who possessed only small flocks seem to have been able to use the public pastures without paying much in fees.[41] Meat and other byproducts, furthermore, remained in high demand from armies and the growing population of Rome. Beef, mutton, and pork were probably more reliable meat products, less susceptible to shifts in culinary fashion than, for example, dormice.

Also worth mentioning is the renewed colonization activities in Italy undertaken by Sulla and later Caesar and Octavian. Though settled on land confiscated from their enemies, these new farmers would have increased demand for agricultural equipment. Cicero suggests that many of Sulla's veterans lived beyond their means and were in serious financial difficulties by 63 BCE (*Cat.* 17ff.). It is also likely that many of them, resettled in unfamiliar regions, were not able to farm as effectively as possible given the considerable regional variation in Italy. Tensions with the remnants of the earlier population would have limited veterans' ability to form networks and gain knowledge of local conditions and practices (though presumably they would be on good terms with their fellow settlers).[42]

Despite the many challenges farmers faced in the late Republic, we should not overstate their difficulties. The proscriptions, at least, mainly affected the wealthy and they were nevertheless able and willing to launch a boom in villa construction.[43]

6.3.3 The early Empire

The imperial period finally brought relative peace and stability to the Italian peninsula and its farmers. While there was fighting in northern Italy during the civil wars of 69 CE and the Marcomanni and Quadi penetrated the same region in 170 CE, by and large the countryside enjoyed safe conditions (aside from occasional low-level brigandage).[44] Nevertheless, our major literary sources appear highly dissatisfied. Pliny the Elder complains about lower yields (*HN* 18.21) and says that *latifundia* have ruined Italy (*HN* 18.35) while Columella voices similar concerns (*Rust.* 1.pr.). Historians have been more reluctant, however, to identify a crisis in Italian agriculture in the early Empire.[45] Aside from the eruption of Vesuvius in 79 CE, there are few signs of major disruption prior to the Antonine Plague.

After a famine in 23 BCE, Augustus took control of the grain supply and subsequent emperors, most notably Claudius and Trajan, helped ensure a (usually) reliable supply from Sicily, North Africa, and Egypt.[46] The disincentives to Italian grain cultivation for the market in areas within easy reach of Rome continued. Olive oil imports from North Africa and Spain also grew, encouraging Roman farmers to devote more attention to other crops.[47] Competition from Spanish oil and Gallic wine may have led to the abandonment of once profitable farms such as those in the hinterland of Luni by the middle of the first century CE.[48]

In 33 CE debt proceedings triggered a financial crisis in which many people lost fortunes due to a sudden drop in real estate prices at a time when they desperately needed to raise cash. Tiberius was eventually able to resolve the crisis which was not agricultural *per se* and apparently had no long-term effects (Tac., *Ann.* 6.16–17). The eruption of Vesuvius was probably the biggest disaster of the first century CE, seriously damaging a wealthy and highly productive region. It seems to have significantly reduced Italian wine production, leading ultimately to overproduction by the end of the first century.[49] Suetonius reports an attempt to regulate the cultivation of grain and vines under Domitian (*Dom.* 7.2). An abundance of wine coupled with a grain shortage led the emperor to impose a (soon abandoned) ban on new vineyards in Italy and a 50 percent reduction in provincial ones. This indicates viticulture was expanding and that Rome relied to some non-negligible extent on Italian grain.

The growth of the imperial *patrimonium* constitutes another major development of this period. To what extent did the emperor's landholdings shape the agricultural economy in Italy, perhaps putting other landowners at a disadvantage? Marco Maiuro cautions against overestimating the extent of the emperor's property but suggests the emperors played a major role in the production and distribution of wine, wool, and wood as well as in the formation of tastes.[50]

There are other scattered signs of trouble. Pliny the Younger complains of the difficulty of finding good tenants (*Ep.* 3.19 and 7.30). Tacitus, in recording the failure of veteran settlement efforts at Tarentum and Antium in 60 CE, notes the depopulation of those places (*Ann.* 14.27).[51] These could easily be regional difficulties and not indications of broader problems. The distribution of amphorae suggests there was a decline in exports from Italy to the provinces but it is also possible that the trade was largely redirected towards Rome and Italian markets not requiring long-distance transport containers (the switch to barrels may also be a factor).[52]

In the first century CE, Rome's provinces were apparently becoming more cost-effective locations for intensive agricultural production thanks to relatively cheap labor.[53] A probable decline in the slave supply would have made agricultural slaves more expensive.[54] Italian agriculture perhaps came to focus somewhat more on extensive rather than intensive cultivation.[55] This would entail a decline in rural investment. Sharecropping may also have grown in importance.[56] In a letter to Maecilius Nepos, Pliny the Younger states that real estate prices had recently risen around Rome because the emperor had required candidates to invest a third of their capital in Italian as opposed to provincial land (*Ep.* 6.19). The implication is that many wealthy Roman citizens did not have large real estate holdings in Italy. Though Pliny does not connect this phenomenon to the profitability of land, it may reflect the relative economic attractiveness of the provinces. Furthermore, it is quite possible that, as Witcher suggests, by this time "the social and political importance of owning an appropriate rural retreat close to Rome exerted such a powerful influence that parts of the countryside began to consume more than they produced."[57] Perhaps these developments led Columella and Pliny the Elder to despair of the state of Italian agriculture. Similar concerns, some have suggested, may have prompted the *alimenta* schemes, initiated by Nerva and expanded in the second century throughout the Italian peninsula, which loaned money to landowners at below the prevailing interest rates in order to fund a food allowance for (some) Italian children.[58] Whether it was intended or not, the loans may have encouraged investment in Italian agriculture. But was that necessary? Despite some provincial competition, the early Empire still offered Italian farmers excellent conditions: peace, accessible markets, and eager buyers. Only the mortality of the Antonine Plague and subsequent political instability (as well as, perhaps, the end of the Roman Warm Period) are likely to have radically altered Italy's situation, dramatically reducing the labor supply as well as urban demand and changing public and private spending habits.

Notes

1　It also plays into persistent fantasies about rural life that are not exclusive to the ancient world.

2　Private alimentary schemes do not seem to have distributed money frequently enough nor to have been generous enough to fully support an individual, but they could have made a significant difference for a poorer recipient with a limited but critical need for cash. However, Woolf (1990) is rightly skeptical of the idea that alimentary schemes were intended to help the poor. See also Liu 2017, 46–8.

3　The purpose of the *alimenta* is debated but few scholars think it was *intended* to support farming even though the quasi-loans to landowners, used to fund the programs, may well have, in some cases, facilitated agricultural investments. Suetonius (*Aug.* 42) mentions cryptically that Augustus took thought for Roman farmers in his management of the *frumentationes*. For efforts to drain the Fucine lake in the early Empire, see Leveau 1993.

4　An exception would be the activities of the Gracchan land commission when it redistributed some of the public land being used by wealthy farmers while providing securer title to their remaining *iugera*.

5　Petronius depicts Trimalchio as having such vast wealth that seventy children are born on just one of his estates in one day (*Sat.* 53).

6 See Rosenstein (2009) for non-economic motives for landownership. With respect to lucrative non-agricultural strategies, Plutarch (*Cat.* 21) notes Cato the Elder's interest in, among other things, investing in overseas trade. Tan (2017) examines some of the ways the Roman elite of the late Republic profited from the provinces.

7 Landownership: land tenure is, of course, more complicated than a simple binary of ownership or lack of ownership. Certain kinds of public land (*ager publicus*) on which rent was, at least in theory, to be paid, provided more or fewer rights to their current occupants. See discussion in Roselaar 2010, 86–145.

8 Most shepherds seem to have been slaves. See Roselaar 2010, 174–5.

9 There *were* distributions elsewhere but not on the same scale. See Woolf 1990, 208.

10 Regional variation: Launaro's analysis of field survey data (2011, 161), for example, suggests the growth of the rural free population in most areas of Italy between the second century BCE and first century CE except in Etruria and parts of Apulia and Calabria. Maiuro (2017, 131) suggests "the unexpected and counter-intuitive hypothesis that Cisalpine Italy had at the same time comparatively low levels of urbanization and a sophisticated economic life."

11 On the potential of modeling agricultural landscapes see, for example, Goodchild and Witcher 2009.

12 Taxation: Launaro (2015, 182) plausibly suggests that the exemption from *tributum* made land a more attractive investment.

13 Slave population: Kay (2014, 178–82) estimates that the slave population grew by 158 percent between 200 and 150 BCE.

14 Settlement following the Second Punic War: Scheidel (2004, 12) estimates forty to sixty-five thousand Roman settlers in the period between 200 and 177 BCE.

15 Available public land: Roselaar 2010, 149–53.

16 Sumptuary laws: Zanda 2011.

17 Effectiveness of sumptuary laws: MacKinnon (2004, 227) says that the zooarchaeological evidence suggests "the sumptuary laws were generally ineffective."

18 Regulating the grain supply: Erdkamp 2005, 240–1.

19 Roads: Witcher (2017, 41). Hitchner (2012, 222) notes the difficulty in "establishing a clear linkage between transport infrastructure and ... economic performance."

20 Increased rural monetization: Hollander 2007, 149–50.

21 Loans: Frayn (1979, 79) argues that the transition from calculating loans of, for example, grain in coin instead of in kind would tend to greatly increase the amount peasants would have to repay.

22 Hellenistic and Punic influences: the Latin translation of Mago's agricultural writings may date to the mid-second century BCE. See White 1970, 18 and Kay 2014, 147–8 and 187.

23 Population boom: de Ligt 2004, 725 and de Ligt 2006, 591.

24 Partible inheritance: Garnsey and Saller 1987, 142–3.

25 Poverty: de Ligt 2006, 603.

26 Slave-run estates: Marzano 2007, 130; Roselaar 2008, 587; and de Ligt 2009.

27 Military service: Rosenstein 2004, 25 and 168.

28 Minimum property qualification: Launaro 2011, 177.

29 Property qualifications: Morstein-Marx and Rosenstein (2006, 631–4) suggest Marius' recruiting decision may not have been as important as is sometimes thought.

30 Supporters of the Gracchi: Rosenstein 2004, 155.

31 Decline in building projects: Tan 2017, 30.

32 Activities of the Gracchan land commission: Roselaar (2010, 252) argues that the distributions were "quite impressive" and provided land for as many as fifteen thousand new farmers between 133 and 129 BCE. However, Scheidel (2004, 12)

believes that the "extent of the initial Gracchan land distribution is impossible to ascertain but unlikely to have been huge."

33 Late second century agrarian laws: Roselaar 2010, 256–78; Kay 2014, 184–5.

34 Wine trade: Paterson 1982, 152; Tchernia 1983, 91–2; McCann et al. 1987, 176; Carandini 1989, 16–17; Arthur 1995, 241; Lo Cascio 2007, 646; Morel 2007, 506; and Rosenstein 2008, 17.

35 *Frumentationes:* Cato the Younger expanded the number of recipients in 63/2 BCE (Plut., *Cat. Min.* 26.1) and P. Clodius started the distribution of free grain in 58 BCE. See also Erdkamp 2005, 241.

36 It is likely that, at least by the early Empire, Rome received more tax grain than was needed for its distributions and sold the surplus to traders and bakers (Erdkamp 2005, 257).

37 Rome's grain supply: Garnsey and Saller (1987, 58) estimate that during the Principate up to 10 percent of the city's supply came from Tuscany, Umbria, Campania, and Apulia.

38 Disruptions of the grain supply: for example, in 57 BCE a shortage led to a special command for Pompey (Plut., *Pomp.* 49–50).

39 Rosenstein 2008, 22.

40 Demand for coinage: Hollander 2016.

41 Public pastures: Roselaar 2010, 134.

42 Veterans as farmers: Hollander 2005.

43 Villa construction: Marzano 2007, 200–1.

44 Brigandage: Columella (*Rust.* 1.pr.7) notes the danger of robbers.

45 Early imperial agricultural crisis: see, for example, Patterson 1987; Lo Cascio 2002, 292ff.

46 Claudius provided incentives to shippers and shipowners involved in the transport of grain. Claudius and Trajan improved harbor facilities near Ostia.

47 Subsidized olive oil at Rome: Harris 2011, 164.

48 Luni: Delano-Smith et al. 1986, 108 and 143.

49 Wine production: Garnsey and Saller 1987, 60.

50 Imperial properties: Maiuro 2012, 145 and 216–29. Alberto Dalla Rosa's PATRI-MONIVM project will eventually offer a new, empire-wide perspective on the role of the emperor's properties in the Roman economy.

51 Depopulation: Tacitus does not attribute the failure to agricultural problems, however, but to the manner in which the veterans were settled (i.e., among strangers and without the presence of leaders with whom they were familiar).

52 Decline in exports: Lo Cascio 2002, 292. Andreau (2015, 110) speaks of an "inversion of the trade patterns" by the end of the first century CE.

53 See, for example, Lo Cascio 2002, 294 and Launaro 2011, 179–80.

54 Reduced slave supply: Lo Cascio 2002, 298 and Andreau 2015, 43.

55 Provincial production: so argues Launaro (2011, 180–3).

56 Increase in sharecropping: so suggests Lo Cascio 2002, 304. See also Marcone 2002, 342–4 and Launaro 2011, 183.

57 Rural retreats: Witcher 2016, 474.

58 *Alimenta:* Marcone (1997, 164–6) discusses the possibility that the *alimenta* promoted agriculture. Woolf (1990, 204) observes that contemporary sources do not explain the rationale behind them. See Launaro (2011, 182) for possible effects of the *alimenta*.

Bibliography

Alcock, S. E., and J. F. Cherry. 2004. *Side-by-Side Survey: Comparative Regional Studies in the Mediterranean World.* Oxford: Oxbow Books.

Aldrete, G. S., S. Bartell, and A. Aldrete. 2013. *Reconstructing Ancient Linen Body Armor: Unraveling the Linothorax Mystery.* Baltimore, MD: Johns Hopkins University Press.

Andreau, J. 1984. "Histoire des métiers bancaires et évolution économique." *Opus* 3: 99–114.

Andreau, J. 2002. "Markets, Fairs and Monetary Loans: Cultural History and Economic History in Roman Italy and Hellenistic Greece." In *Money, Labour and Land: Approaches to the Economies of Ancient Greece*, edited by P. Cartledge, E. Cohen, and L. Foxhall, 113–129. London: Routledge.

Andreau, J. 2015. *The Economy of the Roman World.* Ann Arbor: Michigan Classical Press.

Arthur, P. 1995. "Wine in the West: A View from Campania." In *Italy in Europe: Economic Relationships 700 BC–AD 50,* edited by J. Swaddling, S. Walker, and P. Roberts, 241–251. London: British Museum Press.

Attema, P. 2017. "Landscape Archaeology in Italy: Past Questions, Current State and Future Directions." In *The Economic Integration of Roman Italy: Rural Communities in a Globalizing World*, edited by T. de Haas and G. W. Tol, 426–435. Leiden: Brill.

Aubert, J.-J. 1994. *Business Managers in Ancient Rome: A Social and Economic Study of Institores, 200 BC–AD 250.* Leiden: Brill.

Baldassarre, I. 1990. "Tomba della Mietitura." *Bolletino di archeologia* 5–6: 90–106.

Bang, P. F. 2007. "Trade and Empire – In Search of Organizing Concepts for the Roman Economy." *Past and Present* 195: 3–54.

Bannon, C. 2009. *Gardens and Neighbors: Private Water Rights in Roman Italy.* Ann Arbor: University of Michigan Press.

Barker, G. 1989. "The Archaeology of the Italian Shepherd." *Proceedings of the Cambridge Philological Society* 35: 1–19.

Beard, M. 1998. "Imaginary Horti: Or Up the Garden Path." In *Horti romani: atti del Convegno internazionale, Roma, 4–6 maggio 1995,* edited by M. Cima and E. La Rocca, 23–32. Rome: L'Erma di Bretschneider.

Becker, J. A., and N. Terrenato, eds. 2012. *Roman Republican Villas: Architecture, Context, and Ideology.* Ann Arbor: University of Michigan Press.

Ben Abed, A. 2006. *Tunisian Mosaics: Treasures from Roman Africa, Conservation and Cultural Heritage.* Los Angeles: The Getty Conservation Institute.

Bishop, M. C., and J. C. Coulston. 2006. *Roman Military Equipment: From the Punic Wars to the Fall of Rome.* Oxford: Oxbow.

Blanckenhagen, P. H. von, and C. Alexander. 1962. *The Paintings from Boscotrecase.* Heidelberg: F. H. Kerle Verlag.

Blanton IV, T. R. 2017. "The Economic Functions of Gift Exchange in Pauline Assemblies." In *Paul and Economics: A Handbook*, edited by T. R. Blanton IV and R. Pickett, 279–306. Minneapolis: Fortress Press.

Blok, A. 1969. "South Italian Agro-Towns." *Comparative Studies in Society and History* 2 (2): 121–135.

Bodel, J. 2011. "Slave Labour and Roman Society." In *The Cambridge World History of Slavery*, edited by K. Bradley and P. Cartledge, 311–336. Cambridge: Cambridge University Press.

Bodel, J. 2012. "Villaculture." In *Roman Republican Villas: Architecture, Context, and Ideology,* edited by J. A. Becker and N. Terrenato, 45–60. Ann Arbor: University of Michigan Press.

Bond, S. E. 2016. *Trade and Taboo: Disreputable Professions in the Roman Mediterranean.* Ann Arbor: University of Michigan Press.

Bowes, K., A. M. Mercuri, E. Rattigheri, R. Rinaldi, A. Arnoldus-Huyzendveld, M. Ghisleni, C. Grey, M. MacKinnon, and E. Vaccaro. 2017. "Peasant Agricultural Strategies in Southern Tuscany: Convertible Agriculture and the Importance of Pasture." In *The Economic Integration of Roman Italy: Rural Communities in a Globalizing World*, edited by T. de Haas and G. W. Tol, 170–199. Leiden: Brill.

Bowman, A. K., and A.J.N. Wilson. 2009. "Quantifying the Roman Economy: Integration, Growth, Decline?" In *Quantifying the Roman Economy: Methods and Problems*, edited by A. K. Bowman and A.J.N. Wilson, 1–84. Oxford: Oxford University Press.

Bowman, A. K., and A. Wilson. 2013. *The Roman Agricultural Economy: Organization, Investment, and Production.* Oxford: Oxford University Press.

Bradley, K. R. 1987. "On the Roman Slave Supply and Slavebreeding." *Slavery and Abolition* 8: 42–64.

Bransbourg, G. 2011. "*Fides et Pecunia Numerata.* Chartalism and Metallism in the Roman World: Part 1: The Republic." *American Journal of Numismatics* 23: 87–152.

Braun, T. 1995. "Barley Cakes and Emmer Bread." In *Food in Antiquity*, edited by J. Wilkins, D. Harvey, and M. Dobson, 25–37. Exeter: University of Exeter Press.

Bray, L. 2010. "'Horrible, Speculative, Nasty, Dangerous': Assessing the Value of Roman Iron." *Britannia* 41: 175–185.

Broughton, A. L. 1936. "The Menologia Rustica." *Classical Philology* 31: 353–356.

Brown, A. G., and C. Ellis. 1995. "People, Climate and Alluviation: Theory, Research Design and New Sedimentological and Stratigraphic Data from Etruria." *Papers of the British School at Rome* 63: 45–73.

Brown, V. 1976. "Columella, Lucius Junius Moderatus." *Catalogus Translationum et Commentariorum* 3: 173–193.

Brown, V. 1980a. "Cato, Marcus Porcius." *Catalogus Translationum et Commentariorum* 4: 223–247.

Brown, V. 1980b. "Varro, Marcus Terentius." *Catalogus Translationum et Commentariorum* 4: 451–500.

Brunt, P. A. 1962. "The Army and the Land in the Roman Revolution." *Journal of Roman Studies* 52: 69–86.

Brunt, P. A. 1972. "Review of K. D. White, *Roman Farming.*" *Journal of Roman Studies* 62: 153–158.

Campbell, B. 1996. "Shaping the Rural Environment: Surveyors in Ancient Rome." *Journal of Roman Studies* 86: 74–99.

Campbell, B. 2000. *The Writings of the Roman Land Surveyors. Introduction, Text, Translation and Commentary.* London: Society for the Promotion of Roman Studies.

Carandini, A. 1980. "Il vigneto e la villa del fondo di Settefinestre nel Cosano: un caso di produzione agricola per il mercato transmarino." In *The Seaborne Commerce of Ancient Rome: Studies in Archaeology and History*, edited by J. H. D'Arms and E. C. Kopff, 1–10. Rome: American Academy in Rome.

Carandini, A. 1983. "Columella's Vineyard and the Rationality of the Roman Economy." *Opus* 2: 177–204.

Carandini, A. 1989. "Italian Wine and African Oil: Commerce in a World Empire." In *The Birth of Europe. Archaeology and Social Development in the First Millenium* AD, edited by K. Randsborg, 16–24. Rome: L'Erma di Bretschneider.

Carlà, F. 2013a. "Barley." In *The Encyclopedia of Ancient History*, edited by R. S. Bagnall, K. Brodersen, C. B. Champion, A. Erskine, and S. R. Huebner, 1049–1050. Malden, MA: Blackwell.

Carlà, F. 2013b. "Emmer." In *The Encyclopedia of Ancient History,* edited by R. S. Bagnall, K. Brodersen, C. B. Champion, A. Erskine, and S. R. Huebner, 2389–2390. Malden, MA: Blackwell.

Carlsen, J. 1995. *Vilici and Roman Estate Managers until* AD *284.* Rome: L'Erma di Bretschneider.

Ciaraldi, M. 2005. "How Many Lives Depended on Plants? Specialisation and Agricultural Production at Pompeii." In *Roman Working Lives and Urban Living*, edited by A. Mac Mahon and J. Price, 191–201. Oxford: Oxbow.

Clark, G. 2007. *A Farewell to Alms: A Brief Economic History of the World.* Princeton: Princeton University Press.

Clarysse, W. 2013. "Salt." In *The Encyclopedia of Ancient History,* edited by R. S. Bagnall, K. Brodersen, C. B. Champion, A. Erskine, and S. R. Huebner, 6021–6023. Malden, MA: Blackwell.

Corbier, M. 1989. "The Ambiguous Status of Meat in Ancient Rome." *Food and Foodways* 3 (3): 223–264.

Corretti, A., and M. Benvenuti. 2001. "The Beginning of Iron Metallurgy in Tuscany, with Special Reference to 'Etruria Mineraria'." *Mediterranean Archaeology* 14: 127–145.

Costantini, L., and J. Giorgi. 2009. "The Charred Plant Remains." In *Excavations in the Area Sacra of Vesta (1987–1996)*, edited by R. T. Scott, 125–152. Ann Arbor, Michigan: University of Michigan Press.

Cotton, M. A. 1979. *The Late Republican Villa at Posto, Francolise.* London: British School at Rome.

Cotton, M.A., and G.P.R. Métraux. 1985. *The San Rocco Villa at Francolise.* London and New York: British School at Rome, Institute of Fine Arts of New York University.

Crane, E. 1999. *The World History of Beekeeping and Honey Hunting.* New York: Routledge.

Crawford, M. H. 1974. *Roman Republican Coinage.* Cambridge: Cambridge University Press.

Curti, E., E. Dench, and J. R. Patterson. 1996. "The Archaeology of Central and Southern Roman Italy: Recent Trends and Approaches." *Journal of Roman Studies* 86: 170–189.

Curtis, R. I. 2001. *Ancient Food Technology*. Leiden: Brill.

Curtis, R. I. 2008. "Food Processing and Preparation." In *The Oxford Handbook of Engineering and Technology in the Classical World*, edited by J. P. Oleson, 369–392. Oxford: Oxford University Press.

D'Arms, J. H. 1981. *Commerce and Social Standing in Ancient Rome*. Cambridge, MA: Harvard University Press.

D'Arms, J. H. 2000. "Memory, Money and Status at Misenum: Three New Inscriptions from the *Collegium* of the Augustales." *Journal of Roman Studies* 90: 126–144.

Dalby, A., trans. 1998. *On Farming*. Blackawton: Prospect Books.

Dawson, C. M. 1944. *Romano-Campanian Mythological Landscape Painting*. New Haven: Yale University Press.

De Cecco, M. 1985. "Monetary Theory and Roman History." *Journal of Economic History* 45: 809–822.

de Haas, T. 2017. "The Geography of Roman Italy and Its Implications for the Development of Rural Economies." In *The Economic Integration of Roman Italy: Rural Communities in a Globalizing World*, edited by T. de Haas and G. W. Tol, 51–82. Leiden: Brill.

de Haas, T., and G. W. Tol, eds. 2017. *The Economic Integration of Roman Italy: Rural Communities in a Globalizing World*. Leiden: Brill.

de Ligt, L. 1990. "Demand, Supply, Distribution. The Roman Peasantry between Town and Coutryside. I: Rural Monetisation and Peasant Demand." *Münstersche Beiträge zur antiken Handelsgeschichte* 9: 24–56.

de Ligt, L. 1991. "Demand, Supply, Distribution. The Roman Peasantry between Town and Coutryside. II: Supply, Distribution and a Comparative Perspective." *Münstersche Beiträge zur antiken Handelsgeschichte* 10 (1): 33–77.

de Ligt, L. 1993. *Fairs and Markets in the Roman Empire: Economic and Social Aspects of Periodic Trade in a Pre-Industrial Society*. Amsterdam: J. C. Gieben.

de Ligt, L. 2000. "Studies in Legal and Agrarian History II: Tenancy Under the Republic." *Athenaeum* 88: 377–391.

de Ligt, L. 2004. "Poverty and Demography: The Case of the Gracchan Land Reforms." *Mnemosyne* 57 (6): 725–757.

de Ligt, L. 2006. "The Economy: Agrarian Change During the Second Century." In *A Companion to the Roman Republic*, edited by N. S. Rosenstein and R. Morstein-Marx, 590–605. Oxford: Blackwell.

de Ligt, L. 2009. "Prolegomena for a Low-Count Model of Italy's Agrarian History in the Second Century BC." In *Agricoltura e scambi nell'Italia tardo-repubblicana*, edited by J. Carlsen and E. Lo Cascio, 259–280. Bari: Edipuglia.

de Neeve, P. W. 1984a. *Colonus: Private Farm-Tenancy in Roman Italy During the Republic and Early Principate*. Amsterdam: J. C. Gieben.

de Neeve, P. W. 1984b. *Peasants in Peril: Location and Economy in Italy in the Second Century BC*. Amsterdam: J.C. Gieben.

de Neeve, P. W. 1985. "The Price of Agricultural Land in Roman Italy and the Problem of Economic Rationalism." *Opus* 4: 77–109.

De Sena, E. 2013. "Settefinestre." In *The Encyclopedia of Ancient History*, edited by R. S. Bagnall, K. Brodersen, C. B. Champion, A. Erskine, and S. R. Huebner, 6179. Malden, MA: Blackwell.

Decker, M. 2009. *Tilling the Hateful Earth: Agricultural Production and Trade in the Late Antique East*. Oxford: Oxford University Press.

Degrassi, A. 1960. "Nerva funeraticium plebi urbanae instituit." *Bullettino dell'Istituto di diritto romano* 63: 233–238.

Delano-Smith, C., D. Gadd, N. Mills, and B. Ward-Perkins. 1986. "Luni and the Ager Lunensis: The Rise and Fall of a Roman Town and Its Territory." *Papers of the British School at Rome* 54: 81–146.

Dolansky, F. 2011. "Honouring the Family Dead on the Parentalia: Ceremony, Spectacle, and Memory." *Phoenix* 65: 125–157.

Doody, A. 2007. "Virgil the Farmer? Critiques of the *Georgics* in Columella and Pliny." *Classical Philology* 102 (2): 180–197.

Dunbabin, K. 1978. *The Mosaics of Roman North Africa*. Oxford: Clarendon Press.

Duncan-Jones, R. P. 1962. "Costs, Outlays, and Summae Honorariae from Roman Africa." *Papers of the British School at Rome* 30: 46–115.

Duncan-Jones, R. P. 1974. *The Economy of the Roman Empire: Quantitative Studies*. Cambridge: Cambridge University Press.

Duncan-Jones, R. P. 1982. *The Economy of the Roman Empire: Quantitative Studies*. Second edition. Cambridge: Cambridge University Press.

Dyson, S. 1985. "The Villa of Buccino and the Consumer Model of Roman Rural Development." In *Papers in Italian Archaeology IV*, edited by C. Malone and S. Stoddart, 67–84. Oxford: B.A.R.

Dyson, S. 2003. *The Roman Countryside*. London: Duckworth.

Dyson, S. 2012. "Concluding Remarks." In *Roman Republican Villas: Architecture, Context, and Ideology*, edited by J. A. Becker and N. Terrenato, 129–136. Ann Arbor: University of Michigan Press.

Erdkamp, P. 1999. "Agriculture, Underemployment, and the Cost of Rural Labour in the Roman World." *Classical Quarterly* 49 (2): 556–572.

Erdkamp, P. 2005. *The Grain Market in the Roman Empire: A Social, Political and Economic Study*. Cambridge: Cambridge University Press.

Evans, J. K. 1980. "Plebs Rustica. The Peasantry of Classical Italy." *American Journal of Ancient History* 5: 19–47; 134–173.

Evans, J. K. 1981. "Wheat Production and Its Social Consequences in the Roman World." *Classical Quarterly* 31: 428–442.

Feinman, G. M. 2017. "Roman Economic Practice across Time and Space: An Outside Perspective." In *The Economic Integration of Roman Italy: Rural Communities in a Globalizing World*, edited by T. de Haas and G. W. Tol, 417–425. Leiden: Brill.

Fentress, E. 2005. "Toynbee's legacy: Southern Italy after Hannibal." *Journal of Roman Archaeology* 18: 482–488.

Fentress, E. 2009. "Peopling the Countryside: Roman Demography in the Albegna Valley and Jerba." In *Quantifying the Roman Economy: Methods and Problems*, edited by A. K. Bowman and A.J.N. Wilson, 127–161. Oxford: Oxford University Press.

Finley, M. I. 1976. "Private Farm Tenancy in Italy before Diocletian." In *Studies in Roman Property*, edited by M. I. Finley, 103–121. Cambridge: Cambridge University Press.

Flohr, M. 2013a. *The World of the Fullo: Work, Economy, and Society in Roman Italy*. Oxford: Oxford University Press.

Flohr, M. 2013b. "The Textile Economy of Pompeii." *Journal of Roman Archaeology* 26: 53–78.

Flohr, M. 2013c. "Leather, Leatherwork." In *The Encyclopedia of Ancient History*, edited by R. S. Bagnall, K. Brodersen, C. B. Champion, A. Erskine, and S. R. Huebner, 3984–3987. Malden, MA: Blackwell.

Forbes, H., and L. Foxhall. 1995. "Ethnoarchaeology and Storage in the Ancient Mediterranean: Beyond Risk and Survival." In *Food in Antiquity*, edited by J. Wilkins, D. Harvey, and M. Dobson, 69–86. Exeter: University of Exeter Press.

Forni, G. 2002. "Colture, lavori, tecniche, rendimenti." In *Storia dell'agricoltura italiana I. L'età antica. 2. Italia romana*, edited by G. Forni and A. Marcone, 63–156. Firenze: Edizioni Polistampa.

Forsyth, P. Y. 1988. "In the Wake of Etna, 44 BC" *Classical Antiquity* 7 (1): 49–57.

Foxhall, L. 1990. "The Dependent Tenant: Land Leasing and Labour in Italy and Greece." *Journal of Roman Studies* 80: 97–114.

Foxhall, L. 2007. *Olive Cultivation in Ancient Greece: Seeking the Ancient Economy.* Oxford: Oxford University Press.

Foxhall, L., and H. A. Forbes. 1982. "Sitometreia: The Role of Grain as a Staple Food in Classical Antiquity." *Chiron* 12: 41–90.

Frayn, J. M. 1975. "Wild and Cultivated Plants: A Note on the Peasant Economy of Roman Italy." *Journal of Roman Studies* 65: 32–39.

Frayn, J. M. 1979. *Subsistence Farming in Roman Italy.* London: Centaur Press.

Frayn, J. M. 1984. *Sheep-Rearing and the Wool Trade in Italy During the Roman Period.* Liverpool: F. Cairns.

Frayn, J. M. 1993. *Markets and Fairs in Roman Italy: Their Social and Economic Importance from the Second Century BC to the Third Century AD.* Oxford: Clarendon Press.

Frederiksen, M. W. 1959. "Republican Capua: A Social and Economic Study." *Papers of the British School at Rome* 14: 80–130.

Frederiksen, M. W. 1970. "The Contribution of Archaeology to the Agrarian Problem in the Gracchan Period." *Dialoghi di Archeologia* 4–5: 330–367.

Frier, B. W. 1989–1990. "Law, Economics, and Disasters Down on the Farm: 'Remissio Mercedis' Revisited." *Bullettino dell'Istituto di diritto romano* 31–32: 237–270.

Frier, B. W., and D. P. Kehoe. 2007. "Law and Economic Institutions." In *The Cambridge Economic History of the Greco-Roman World*, edited by W. Scheidel, I. Morris, and R. P. Saller, 113–143. Cambridge: Cambridge University Press.

Gaastra, J. S. 2016. "Sheep." In *The Encyclopedia of Ancient History*, edited by R. S. Bagnall, K. Brodersen, C. B. Champion, A. Erskine, and S. R. Huebner. Malden, MA: Blackwell.

Gabba, E., and F. Coarelli. 1975. "Mercati e fiere nell'Italia romana." *Studi Classici e Orientali* 24: 141–166.

Garnsey, P.D.A. 1968. "Trajan's Alimenta: Some Problems." *Historia: Zeitschrift für alte Geschichte* 17 (3): 367–381.

Garnsey, P.D.A. 1979. "Where Did Italian Peasants Live?" *Proceedings of the Cambridge Philological Society* 205: 1–25.

Garnsey, P.D.A. 1988. *Famine and Food Supply in the Graeco-Roman World. Responses to Risk and Crisis.* Cambridge: Cambridge University Press.

Garnsey, P.D.A. 1991. "Mass Diet and Nutrition in the City of Rome." In *Nourrir la plebe: actes du colloque tenu à Genève les 28 et 29. IX. 1989 en hommage à Denis van Berchem*, edited by A. Giovannini, 67–99. Basel: F. Reinhardt.

Garnsey, P.D.A. 1999. *Food and Society in Classical Antiquity.* Cambridge: Cambridge University Press.

Garnsey, P.D.A., and R. P. Saller. 1987. *The Roman Empire: Economy, Society and Culture*. Berkeley and Los Angeles: University of California Press.

Garnsey, P.D.A., and R. P. Saller. 2015. *The Roman Empire: Economy, Society and Culture*. Second edition. Oakland: University of California Press.

Garnsey, P D A., and G. Woolf. 1989. "Patronage of the Rural Poor in the Roman World." In *Patronage in Ancient Society*, edited by A. Wallace-Hadrill, 153–170. London: Routledge.

Ghisleni, M., E. Vaccaro, K. Bowes, A. Arnoldus, M. MacKinnon, and F. Marani. 2011. "Excavating the Roman Peasant I: Excavations at Pievina (GR)." *Papers of the British School at Rome* 79: 95–145.

Giovannini, A. 1985. "Le sel et la fortune de Rome." *Athenaeum* 63: 373–387.

Gleba, M. 2004. "Linen Production in Pre-Roman and Roman Italy." In *Purpureae Vestes. Actas del I Symposium Internacional sobre Textiles y Tintes del Mediterráneo en época romana (Ibiza, 8 al 10 de noviembre, 2002)*, edited by C. Alfaro-Giner, J. P. Wild, and B. Costa, 29–37. Valencia: Universitat de València.

Goodchild, H. 2013. "Agriculture and the Environment of Republican Italy." In *A Companion to the Archaeology of the Roman Republic*, edited by J. DeRose Evans, 198–213. Malden, MA: Wiley-Blackwell.

Goodchild, H., and R. Witcher. 2009. "Modelling the Agricultural Landscapes of Republican Italy." In *Agricoltura e scambi nell'Italia tardo-repubblicana*, edited by J. Carlsen and E. Lo Cascio, 187–220. Bari: Edipuglia.

Green, C.M.C. 2007. *Roman Religion and the Cult of Diana at Aricia*. Cambridge: Cambridge University Press.

Green, C.M.C. 2012. "The Shepherd of the People: Varro on Herding for the Villa Publica in *De re rustica* 2." In *Roman Republican Villas: Architecture, Context, and Ideology*, edited by J. A. Becker and N. Terrenato, 32–44. Ann Arbor: University of Michigan Press.

Greene, K. 1986. *The Archaeology of the Roman Economy*. London: B. T. Batsford Ltd.

Groot, M. 2016. *Livestock for Sale. Animal Husbandry in a Roman Frontier Zone. The Case Study of the Civitas Batavorum, Amsterdam Achaeological Studies*. Amsterdam: Amsterdam University Press.

Haarer, P. 2013. "Iron." In *The Encyclopedia of Ancient History*, edited by R. S. Bagnall, K. Brodersen, C. B. Champion, A. Erskine, and S. R Huebner, 3498–3500. Malden, MA: Blackwell.

Halstead, P. 1987. "Traditional and Ancient Rural Economies in Mediterranean Europe: Plus ça change?" *Journal of Hellenistic Studies* 107: 77–87.

Hanelt, P., and Institute of Plant Genetics and Crop Plant Research, eds. 2001. *Mansfeld's Encyclopedia of Agricultural and Horticultural Crops (Except Ornamentals)*. Gatersleben, Germany: Institut für Pflanzengenetik und Kulturpflanzenforschung.

Harris, W. V. 1971. *Rome in Etruria and Umbria*. Oxford: Oxford University Press.

Harris, W. V. 1993. "Between Archaic and Modern: Problems in Roman Economic History." In *The Inscribed Economy: Production and Distribution in the Roman Empire in the Light of Instrumentum Domesticum: The Proceedings of a Conference Held at the American Academy in Rome on 10–11 January, 1992*, edited by W. V. Harris, 11–29. Ann Arbor: University of Michigan.

Harris, W. V. 1999. "Demography, Geography and the Sources of Roman Slaves." *Journal of Roman Studies* 98: 62–75.

Harris, W. V. 2005. "The Mediterranean and Ancient History." In *Rethinking the Mediterranean*, edited by W. V. Harris, 1–42. Oxford: Oxford University Press.

Harris, W. V. 2011. "Trade [70–192 AD]." In *Rome's Imperial Economy: Twelve Essays*, edited by W. V. Harris, 155–187. Oxford: Oxford University Press.

Harris, W. V. 2013. "Defining and Detecting Mediterranean Deforestation, 800 BCE to 700 CE." In *The Ancient Mediterranean Environment between Science and History*, edited by W. V. Harris, 173–194. Brill: Leiden.

Harvey, S. M. 2010. "Iron Tools from a Roman Villa at Boscoreale, Italy, in the Field Museum and the Kelsey Museum of Archaeology." *American Journal of Archaeology* 114: 697–714.

Healy, J. F., trans. 1991. *Natural History: A Selection*. Oxford: Oxford University Press.

Heinrich, F. 2017. "Modelling Crop-Selection in Roman Italy. The Economics of Agricultural Decision Making in a Globalizing Economy." In *The Economic Integration of Roman Italy: Rural Communities in a Globalizing World*, edited by T. de Haas and G. W. Tol, 141–169. Leiden: Brill.

Henderson, J. 2002. "Columella's Living Hedge: The Roman Gardening Book." *Journal of Roman Studies* 92: 110–133.

Higginbotham, J. 1997. *Piscinae: Artificial Fishponds in Roman Italy*. Chapel Hill: University of North Carolina Press.

Hin, S. 2013. *The Demography of Roman Italy. Population Dynamics in an Ancient Conquest Society (201 BCE–CE 14)*. Cambridge: Cambridge University Press.

Hitchner, R. B. 2002. "Olive Production and the Roman economy: The Case for Intensive Growth in the Roman Empire." In *The Ancient Economy*, edited by W. Scheidel and S. Von Reden, 71–83. New York: Routledge.

Hitchner, R. B. 2012. "Roads, Integration, Connectivity, and Economic Performance in the Roman Empire." In *Highways, Byways, and Road Systems in the Pre-Modern World*, edited by S. E. Alcock, J. Bodel, and R.J.A. Talbert, 222–234. New York: Wiley Blackwell.

Hollander, D. B. 2005. "Veterans, Agriculture, and Monetization in the Late Roman Republic." In *A Tall Order: Writing the Social History of the Ancient World. Essays in Honor of William V. Harris*, edited by Z. Varhelyi and J.-J. Aubert, 229–239. Leipzig: Teubner.

Hollander, D. B. 2007. *Money in the Late Roman Republic*. Leiden: Brill.

Hollander, D. B. 2015. "Risky Business: Traders in the Roman World." In *Traders in the Ancient Mediterranean*, edited by T. Howe, 141–172. Chicago: Ares.

Hollander, D. B. 2016. "Lawyers, Friends, and Money: Portfolios of Power in the Late Republic." In *Money and Power in the Late Republic*, edited by H. Beck, M. Jehne, and J. Serrati, 18–25. Brussels: Latomus.

Hollander, D. B. 2017. "The Roman Economy in the Early Empire: An Overview." In *Paul and Economics: A Handbook*, edited by T. R. Blanton IV and R. Pickett, 1–22. Minneapolis: Fortress Press.

Holleran, C. 2012. *Shopping in Ancient Rome: The Retail Trade in the Late Republic and the Principate*. Oxford: Oxford University Press.

Hopkins, K. 1980. "Taxes and Trade in the Roman Empire (200 BC–AD 400)." *Journal of Roman Studies* 70: 101–125.

Hopkins, K. 1995/1996. "Rome, Taxes, Rents and Trade." *Kodai* 6/7: 41–75.

Hopkins, K. 2000. "Rents, Taxes, Trade and the City of Rome." In *Mercati permanenti e mercati periodici nel mondo romano: Atti degli Incontri capresi di storia dell'economia antica (Capri 13–15 ottobre 1997)*, edited by E. Lo Cascio, 253–267. Bari: Edipuglia.

Horden, P. 2013. "Climate." In *The Encyclopedia of Ancient History,* edited by R. S. Bagnall, K. Brodersen, C. B. Champion, A. Erskine, and S. R. Huebner, 1582–1583. Malden, MA: Blackwell.

Horden, P., and N. Purcell. 2000. *The Corrupting Sea: A Study of Mediterranean History.* Oxford: Blackwell.

Horsfall, N. 2001. "The Unity of Roman Italy: Anomalies in Context." *Scripta Classica Israelica* 20: 39–50.

Jackson, R. 2005. "The Role of Doctors in the City." In *Roman Working Lives and Urban Living,* edited by A. MacMahon and J. Price, 202–220. Oxford: Oxbow.

Jashemski, W. F. 1979. *The Gardens of Pompeii: Herculaneum and the Villas Destroyed by Vesuvius.* New Rochelle, NY: Caratzas Brothers.

Jashemski, W. F., and F. G. Meyer. 2002. *The Natural History of Pompeii.* New York: Cambridge University Press.

Jashemski, W. F., F. G. Meyer, and M. Ricciardi. 2002. "Catalogue of Plants." In *The Natural History of Pompeii,* edited by W. F. Jashemski and F. G. Meyer, 84–180. Cambridge: Cambridge University Press.

Jasny, N. 1942. "Competition Among Grains in Classical Antiquity." *American Historical Review* 47 (4): 747–764.

Johnson, A. W., and T. Earle. 2000. *The Evolution of Human Societies.* Second edition. Stanford: Stanford University Press.

Johnston, D. 1999. *Roman Law in Context.* Cambridge: Cambridge University Press.

Jongman, W. 1988. "Adding It Up." In *Pastoral Economies in Classical Antiquity,* edited by C. R. Whittaker, 210–212. Cambridge: Cambridge Philological Society.

Jongman, W. 2000. "Wool and the Textile Industry of Roman Italy: A Working Hypothesis." In *Mercati permanenti e mercati periodici nel mondo romano: Atti degli Incontri capresi di storia dell'economia antica (Capri 13–15 ottobre 1997),* edited by E. Lo Cascio, 187–197. Bari: Edipuglia.

Jongman, W. 2003a. "Slavery and the Growth of Rome. The Transformation of Italy in the Second and First Centuries BCE." In *Rome the Cosmopolis,* edited by C. Edwards and G. Woolf, 100–122. Cambridge: Cambridge University Press.

Jongman, W. 2003b. "A Golden Age: Death, Money Supply and Social Succession in the Roman Empire." In *Credito e moneta nel mondo romano: Atti degli Incontri capresi di storia dell'economia antica (Capri 12–14 ottobre 2000),* edited by E. Lo Cascio, 181–196. Bari: Edipuglia.

Jongman, W. 2007a. "Gibbon Was Right: The Decline and Fall of the Roman Economy." In *Crises and the Roman Empire: Proceedings of the Seventh Workshop of the International Network Impact of Empire (Nijmegen, June 20–24, 2006),* edited by O. Hekster, G. de Kleijn, and D. Slootjes, 183–199. Leiden: Brill.

Jongman, W. 2007b. "The Early Roman Empire: Consumption." In *The Cambridge Economic History of the Greco-Roman World,* edited by W. Scheidel, I. Morris, and R. P. Saller, 592–618. Cambridge: Cambridge University Press.

Jongman, W. 2017. "The Benefits of Market Integration: Five Centuries of Prosperity in Roman Italy." In *The Economic Integration of Roman Italy: Rural Communities in a Globalizing World,* edited by T. de Haas and G. W. Tol, 15–27. Leiden: Brill.

Kamprath, E. J. and T. J. Smyth. 2005. "Liming." In *Encyclopedia of Soils in the Environment* Vol. 3, edited by D. Hillel, 350–358. Amsterdam: Elsevier.

Kay, P. 2014. *Rome's Economic Revolution.* Oxford: Oxford University Press.

Kehoe, D. P. 1989. "Approaches to Economic Problems in the 'Letters' of Pliny the Younger: The Question of Risk in Agriculture." *Aufstieg und Niedergang der römischen Welt* II (33): 555–590.

Kehoe, D. P. 1993. "Economic Rationalism in Roman Agriculture." *Journal of Roman Archaeology* 6: 476–484.

Kehoe, D. P. 1997. *Investment, Profit and Tenancy: The Jurists and the Roman Agrarian Economy.* Ann Arbor: University of Michigan Press.

Kehoe, D. P. 2007a. "The Early Roman Empire: Production." In *The Cambridge Economic History of the Greco-Roman World*, edited by W. Scheidel, I. Morris, and R. P. Saller, 543–569. Cambridge: Cambridge University Press.

Kehoe, D. P. 2007b. *Law and the Rural Economy in the Roman Empire.* Ann Arbor: University of Michigan.

Kehoe, D. P. 2012. "Contract Labor." In *The Cambridge Companion to the Roman Economy*, edited by W. Scheidel, 114–130. Cambridge: Cambridge University Press.

Kellum, B. A. 1994. "The Construction of Landscape in Augustan Rome: The Garden Room at the Villa ad Gallinas." *The Art Bulletin* 76 (2): 211–224.

Kenney, E. J. 1984. *The Ploughman's Lunch: Moretum: A Poem Ascribed to Virgil.* Bristol: Bristol Classical Press.

Keppie, L. 1983. *Colonisation and Veteran Settlement in Italy: 47–14* BC London: British School at Rome.

Ker, J. 2010. "*Nundinae:* The Culture of the Roman Week." *Phoenix* 64 (3/4): 360–385.

King, H. 2001. *Greek and Roman Medicine.* London: Bristol Classical Press.

Kleijwegt, M. 2002. "Textile Manufacturing for a Religious Market, Artemis and Diana as Tycoons of Industry." In *After the Past: Essays in Ancient History in Honour of H. W. Pleket*, edited by W. Jongman and M. Kleijwegt, 81–134. Leiden: Brill.

Knapp, R. C. 1977. "The Date and Purpose of the Iberian Denarii." *Numismatic Chronicle* 137: 1–18.

Kron, G. 2000. "Roman Ley-Farming." *Journal of Roman Archaeology* 13: 277–287.

Kron, G. 2002. "Archaeozoological Evidence for the Productivity of Roman Livestock Farming." *Münstersche Beiträge zur antiken Handelsgeschichte* 21 (2): 53–73.

Kron, G. 2004. "Roman Livestock Farming in Southern Italy: The Case against Environmental Determinism." In *Espaces intégrés et ressources naturelles dans l'empire romain: Actes du colloque de l'Université de Laval, Québec (5–8 mars 2003)*, edited by M. Clavel-Lévêque and E. Hermon, 119–134. Besançon: Presses Universitaires de Franche-Comte.

Kron, G. 2012. "Food Production." In *The Cambridge Companion to the Roman Economy*, edited by W. Scheidel, 156–174. Cambridge: Cambridge University Press.

Kron, G. 2013. "Agriculture, Roman Empire." In *The Encyclopedia of Ancient History*, edited by R. S. Bagnall, K. Brodersen, C. B. Champion, A. Erskine, and S. R. Huebner, 217–222. Malden, MA: Blackwell.

Kron, G. 2015. "Agriculture." In *A Companion to Food in the Ancient World*, edited by J. Wilkins and R. Nadeau, 160–172. Malden, MA: Wiley Blackwell.

Kron, G. 2017. "The Diversification and Intensification of Italian Agriculture: The Complementary Roles of the Small and Wealthy Farmer." In *The Economic Integration of Roman Italy: Rural Communities in a Globalizing World*, edited by T. de Haas and G. W. Tol, 112–140. Leiden: Brill.

Kronenberg, L. 2009. *Allegories of Farming from Greece and Rome: Philosophical Satire in Xenophon, Varro, and Virgil*. Cambridge: Cambridge University Press.

Lafon, X. 1993. "L'Huile en Italie centrale à l'époque républicaine: une production sous-estimée?" In *La production du vin et de l'huile en Méditerranée*, edited by M.-C. Amouretti and J.-P. Brun, 263–281. Athens and Paris: École Française d'Athènes and Diffusion de Boccard.

Larew, H. G. 2002. "Insects: Evidence from Wall Paintings, Sculpture, Mosaics, Carbonized Remains, and Ancient Authors." In *The Natural History of Pompeii*, edited by W. F. Jashemski and F. G. Meyer, 315–326. Cambridge: Cambridge University Press.

Launaro, A. 2011. *Peasants and Slaves: The Rural Population of Roman Italy (200 BC to AD 100)*. Cambridge: Cambridge University Press.

Launaro, A. 2015. "The Nature of the Villa Economy." In *Ownership and Exploitation of Land and Natural Resources in the Roman World*, edited by P. Erdkamp, K. Verboven, and A. Zuiderhoek, 173–186. Oxford: Oxford University Press.

Launaro, A. 2017. "Something Old, Something New: Social and Economic Developments in the Countryside of Roman Italy between Republic and Empire." In *The Economic Integration of Roman Italy: Rural Communities in a Globalizing World*, edited by T. de Haas and G. W. Tol, 85–111. Leiden: Brill.

Laurence, R. 1998. "Land Transport in Roman Italy: Costs, Practice and the Economy." In *Trade, Traders and the Ancient City*, edited by H. Parkins and C. Smith, 129–148. London: Routledge.

Laurence, R. 1999. *The Roads of Roman Italy: Mobility and Cultural Change*. London: Routledge.

Leach, E. W. 1980. "Sacral-Idyllic Landscape Painting and the Poems of Tibullus' First Book." *Latomus* 39 (1): 47–69.

Leach, E. W. 1988. *The Rhetoric of Space: Literary and Artistic Representations of Landscape in Republican and Augustan Rome*. Princeton: Princeton University Press.

Leveau, P. 1993. "Mentalité économique et grands travaux: le drainage du lac Fucin. Aux origines d'un modèle." *Annales ESC* 48: 3–16.

Ling, R. 1977. "Studius and the Beginnings of Roman Landscape Painting." *Journal of Roman Studies* 67: 1–16.

Ling, R. 1991. *Roman Painting*. Cambridge: Cambridge University Press.

Lintott, A. W. 1986. *Violence in Republican Rome*. Oxford: Oxford University Press.

Lintott, A. W. 1992. *Judicial Reform and Land Reform in the Roman Republic: A New Edition, with Translation and Commentary, of the Laws from Urbino*. Cambridge: Cambridge University Press.

Lipkin, S. 2012. *Textile-making in Central Tyrrhenian Italy from the Final Bronze Age to the Republican Period*. Oxford: Archaeopress.

Lirb, H. J. 1993. "Partners in Agriculture: The Pooling of Resources in Rural *Societates* in Roman Italy." In *De Agricultura: In Memoriam Pieter Willem De Neeve*, edited by H. Sancisi-Weerdenburg, R. J. van der Spek, H. C. Teitler, and H. T. Wallinga, 263–295. Amsterdam: J. C. Gieben.

Liu, J. 2017. "Urban Poverty in the Roman Empire: Material Conditions." In *Paul and Economics: A Handbook*, edited by T. R. Blanton IV and R. Pickett, 23–56. Minneapolis: Fortress Press.

Lo Cascio, E. 1982. "'Obaerarii' ('obaerati'). la nozione della dipendenza in Varrone." *Index* 11: 265–284.

Lo Cascio, E. 2002. "La proprietà della terra, I percettori dei prodotti e della ren-dita." In *Storia dell'agricoltura italiana I. L'età antica. 2. Italia romana*, edited by G. Forni and A. Marcone, 259–314. Firenze: Edizioni Polistampa.

Lo Cascio, E. 2007. "The Early Roman Empire: The State and the Economy." In *The Cambridge Economic History of the Greco-Roman World*, edited by W. Scheidel, I. Morris, and R. P. Saller, 619–647. Cambridge: Cambridge University Press.

Lo Cascio, E., ed. 2000. *Mercati permanenti e mercati periodici nel mondo romano: Atti degli Incontri capresi di storia dell'economia antica (Capri 13–15 ottobre 1997)*. Bari: Edipuglia.

Long, C. R. 1992. "The Pompeii Calendar Medallions." *American Journal of Archaeology* 96 (3): 477–501.

Longenecker, B. W. 2010. *Remember the Poor: Paul, Poverty, and the Greco-Roman World*. Grand Rapids, MI: Eerdmans.

MacKinnon, M. 2004. *Production and Consumption of Animals in Roman Italy: Integrating the Zooarchaeological and Textual Evidence*. Portsmouth, R.I: Journal of Roman Archaeology.

MacMullen, R. 1970. "Roman Market Days." *Phoenix* 24: 333–341.

Maiuri, A. 1953. *Roman Painting*. Lausanne: Skira.

Maiuro, M. 2012. *Res Caesaris: Ricerche sulla proprietà imperiale nel Principato*. Bari: Edipuglia.

Maiuro, M. 2017. "Northern Italy: Urbanization, Demography and Agrarian Output." In *Popolazione e risorse nell'Italia del Nord dalla romanizzazione ai Longobardi*, edited by E. Lo Cascio and M. Maiuro, 99–147. Bari: Edipuglia.

Malik, S. J., and H. Nazli. 1999. "Rural Poverty and Credit Use: Evidence from Pakistan." *The Pakistan Development Review* 38 (4): 699–716.

Manning, S. W. 2013. "The Roman World and Climate: Context, Relevance of Climate Change, and Some Issues." In *The Ancient Mediterranean Environment Between Science and History*, edited by W. V. Harris, 103–170. Leiden: Brill.

Marcone, A. 1997. *Storia dell'agricoltura romana. Dal mondo arcaico all'età imperiale*. Rome: La Nuova Italia Scientifica.

Marcone, A. 2002. "La circolazione dei prodotti." In *Storia dell'agricoltura italiana I. L'età antica. 2. Italia romana*, edited by G. Forni and A. Marcone, 315–352. Firenze: Edizioni Polistampa.

Margaritis, E., and M. K. Jones. 2008. "Greek and Roman Agriculture." In *The Oxford Handbook of Engineering and Technology in the Classical World*, edited by J. P. Oleson, 158–174. Oxford: Oxford University Press.

Maróti, E. 1976. "The Vilicus and the Villa-System in Ancient Italy." *Oikumene* 1: 109–124.

Martin, C. J., T. S. Richter, J. Rowlandson, R. Takahashi, and D. J. Thompson. 2014. "Leases." In *Law and Legal Practice in Egypt from Alexander to the Arab Conquest*, edited by J. G. Keenan, J. G. Manning, and U. Yiftach-Firanko, 339–400. Cambridge: Cambridge University Press.

Martin, R. 1971. *Recherches sur les agronomes latins et leurs conceptions économiques et sociales*. Paris: Les Belles lettres.

Marzano, A. 2007. *Roman Villas in Central Italy. A Social and Economic History*. Leiden: Brill.

Marzano, A. 2009. "'Agros coemendo et colendo in gloriam': Villas and Farms in the Agrarian Economy of the Republic." In *The Italians on the Land: Changing Perspectives on Republican Italy Then and Now*, edited by A. Keaveney and L. Eamshaw-Brown, 31–46. Newcastle upon Tyne: Cambridge Scholars Publishing.

Marzano, A. 2013. *Harvesting the Sea: The Exploitation of Marine Resources in the Roman Mediterranean*. Oxford: Oxford University Press.

Marzano, A. 2015. "The Variety of Villa Production." In *Ownership and Exploitation of Land and Natural Resources in the Roman World*, edited by P. Erdkamp, K. Verboven, and A. Zuiderhoek, 187–206. Oxford: Oxford University Press.

Mattingly, D. J. 1996. "First Fruit? The Olive in the Roman World." In *Human Landscapes in Classical Antiquity: Environment and Culture*, edited by J. Salmon and G. Shipley, 213–253. London: Routledge.

Mattingly, D. J. 2006. "The Imperial Economy." In *A Companion to the Roman Empire*, edited by D. S. Potter, 283–297. Malden, MA: Blackwell.

McCann, A. A., J. Bourgeois, E. K. Gazda, J. P. Oleson, and E. L. Will. 1987. *The Roman Port and Fishery of Cosa: A Center of Ancient Trade*. Princeton: Princeton University Press.

McCormick, A., U. Büntgen, M. A. Cane, E. R. Cook, K. Harper, P. Huybers, T. Litt, S. W. Manning, P. A. Mayewski, A.F.M. More, K. Nicolussi, and W. Tegel. 2012. "Climate Change During and After the Roman Empire: Reconstructing the Past from Scientific and Historical Evidence." *Journal of Interdisciplinary History* 43 (2): 169–220.

McGinn, T.A.J. 2004. *The Economy of Prostitution in the Roman World: A Study of Social History and the Brothel*. Ann Arbor: University of Michigan Press.

Meijer, F. 1990. "The Financial Aspects of the *Leges Frumentariae* of 123–58 BC." *Münstersche Beiträge zur antiken Handelsgeschichte* 9 (2): 14–23.

Mercando, L. 1965. "Falerone (Ascoli Piceno): Rinvenimento di tombe romane." *Notizie degli scavi di antichità* 19: 253–273.

Mercuri, A. M., C. A. Accorsi, and M. Bandini Mazzanti. 2002. "The Long History of Cannabis and its Cultivation by the Romans in Central Italy, Shown by Pollen Records from Lago Albano and Lago di Nemi." *Vegetation History and Archaeobotany* 11: 263–276.

Mishkin, F. S. 1992. *The Economics of Money, Banking, and Financial Markets*. New York: HarperCollins.

Mitchell, S. 2005. "Olive Cultivation in the Economy of Roman Asia Minor." In *Patterns in the Economy of Roman Asia Minor*, edited by S. Mitchell and C. Katsari, 83–114. Swansea: The Classical Press of Wales.

Morel, J.-P. 2007. "Early Rome and Italy." In *The Cambridge Economic History of the Greco-Roman World*, edited by W. Scheidel, I. Morris, and R. P. Saller, 487–510. Cambridge: Cambridge University Press.

Moritz, L. A. 1955. "Husked and 'Naked' Grain." *Classical Quarterly* 5: 129–134.

Morley, N. 1996. *Metropolis and Hinterland: The City of Rome and the Italian Economy, 200 BC–AD 200*. Cambridge: Cambridge University Press.

Morley, N. 2000. "Markets, Marketing and the Roman Élite." In *Mercati permanenti e mercati periodici nel mondo romano: Atti degli Incontri capresi di storia dell'economia antica (Capri 13–15 ottobre 1997)*, edited by E. Lo Cascio, 211–221. Bari: Edipuglia.

Morley, N. 2005. "The Salubriousness of the Roman City." In *Health in Antiquity*, edited by H. King, 192–204. London: Routledge.

Morley, N. 2007a. *Trade in Classical Antiquity*. Cambridge: Cambridge University Press.

Morley, N. 2007b. "The Early Roman Empire: Distribution." In *The Cambridge Economic History of the Greco-Roman World*, edited by W. Scheidel, I. Morris, and R. P. Saller, 570–591. Cambridge: Cambridge University Press.

Morstein-Marx, R., and N. S. Rosenstein. 2006. "The Transformation of the Republic." In *A Companion to the Roman Republic*, edited by N. S. Rosenstein and R. Morstein-Marx, 625–637. Malden, MA: Wiley Blackwell.

Netting, R. McC. 1993. *Smallholders, Householders: Farms Families and the Ecology of Intensive, Sustainable Agriculture*. Stanford: Stanford University Press.

Neudecker, R. 2015. "Flowers in Cult, Rome." In *The Encyclopedia of Ancient History*, edited by R. S. Bagnall, K. Brodersen, C. B. Champion, A. Erskine, and S. R. Huebner. Malden, MA: Blackwell.

Nisbet, C. 1967. "Interest Rates and Imperfect Competition in the Informal Credit Market of Rural Chile." *Economic Development and Cultural Change* 16 (1): 73–90.

Nisbet, C. 1969. "The Relationship between Institutional and Informal Credit Markets in Rural Chile." *Land Economics* 45 (2): 162–173.

Nutton, V. 1985. "The Drug Trade in Antiquity." *The Journal of the Royal Society of Medicine* 78: 138–145.

Olcese, G. 2017. "Wine and Amphorae in Campania in the Hellenistic Age: The Case of Ischia." In *The Economic Integration of Roman Italy: Rural Communities in a Globalizing World*, edited by T. de Haas and G. W. Tol, 299–321. Leiden: Brill.

Orengo, H. A., J. M. Palet, A. Ejarque, Y. Miras, and S. Riera. 2013. "Pitch Production during the Roman Period: An Intensive Mountain Industry for a Globalised Economy." *Antiquity* 87: 802–814.

Parrish, D. 1979. "Two Mosaics from Roman Tunisia: An African Variation of the Season Theme." *American Journal of Archaeology* 83: 279–285.

Pasquinucci, M., and S. Menchelli. 2017. "Rural, Urban and Suburban Communities and Their Economic Interconnectivity in Coastal North Etruria (2nd Century BC–2nd Century AD)." In *The Economic Integration of Roman Italy: Rural Communities in a Globalizing World*, edited by T. de Haas and G. W. Tol, 322–341. Leiden: Brill.

Passi Pitcher, L. 1987. *Sub ascia: Una necropolis romana a Nave*. Modena: Edizioni Panini.

Paterson, J. 1982. "'Salvation from the Sea:' Amphorae and Trade in the Roman World." *Journal of Roman Studies* 72: 146–157.

Paterson, J. 1998. "Trade and Traders in the Roman World: Scale, Structure, and Organisation." In *Trade, Traders and the Ancient City*, edited by H. Parkins and C. Smith, 149–167. London and New York: Routledge.

Patterson, J. R. 1987. "Crisis: What Crisis? Rural Change and Urban Development in Imperial Appennine Italy." *Papers of the British School at Rome* 55: 115–146.

Peacock, D.P.S., and D. F. Williams. 1986. *Amphorae and the Roman Economy: An Introductory Guide*. London: Longman.

Peña, J. T. 2017. "Issues in the Study of Rural Craft Production in Roman Italy." In *The Economic Integration of Roman Italy: Rural Communities in a Globalizing World*, edited by T. de Haas and G. W. Tol, 203–230. Leiden: Brill.

Percival, J. 1976. *The Roman Villa: An Historical Introduction*. London: B. T. Batsford Ltd.

Phillips, E. J. 1973. "The Roman Law on the Demolition of Buildings." *Latomus* 32: 86–95.

Pirazzoli, P. A. 1976. "Sea Level Variations in the Northwest Mediterranean During Roman Times." *Science* 194: 519–521.

Prowse, T., H. P. Schwarcz, S. R. Saunders, R. Macchiarelli, and L. Bondioli. 2010. "Stable Isotope and Mitochrondrial DNA Evidence for Geographic Origins on a Roman Estate at Vagnari (Italy)." In *Roman Diasporas: Archaeological Approaches to Mobility and Diversity in the Roman Empire*, edited by H. Eckhardt, 175–197. Portsmouth, RI: Journal of Roman Archaeology.

Purcell, N. 1985. "Wine and Wealth in Ancient Italy." *Journal of Roman Studies* 75: 1–19.

Purcell, N. 1987. "Tomb and Suburb." In *Römische Gräberstrassen. Selbstdarstellung, Status, Standard: Kolloquium in München vom 28. bis 30. Oktober 1985*, edited by H. von Hesberg and P. Zanker, 24–41. Munich: Verlag der Bayerischen Akademi.

Purcell, N. 1995. "The Roman Villa and the Landscape of Production." In *Urban Society in Roman Italy*, edited by H. K. Lomas and T. J. Cornell, 151–179. London: St. Martin's Press.

Rathbone, D. 1981. "The Development of Agriculture in the 'Ager Cosanus' during the Roman Republic: Problems of Evidence and Interpretation." *Journal of Roman Studies* 71: 10–23.

Rathbone, D. 2008. "Poor Peasants and Silent Sherds." In *People, Land, and Politics: Demographic Developments and the Transformation of Roman Italy 300 BC–AD 14*, edited by L. de Ligt and S. J. Northwood, 305–332. Leiden: Brill.

Rawson, E. 1976. "The Ciceronian Aristocracy and Its Properties." In *Studies in Roman Property*, edited by M. I. Finley, 85–102. Cambridge: Cambridge University Press.

Reay, B. 2005. "Agriculture, Writing, and Cato's Aristocratic Self-Fashioning." *Classical Antiquity* 24: 331–361.

Reay, B. 2012. "Cato's *De agri cultura* and the Spectacle of Expertise." In *Roman Republican Villas: Architecture, Context, and Ideology*, edited by J. A. Becker and N. Terrenato, 61–7. Ann Arbor: University of Michigan Press.

Reekmans, T. 1986. "The Motives of the Roman Farmer's Economic Options." In *Hommages à J. Veremans.*, edited by Fr. Decreus and C. Deroux, 259–273. Bruxelles: Latomus.

Reger, G. 2005. "The Manufacture and Distribution of Perfume." In *Making, Moving and Managing: The New World of Ancient Economies, 323–31 BC*, edited by Z. H. Archibald, J. K. Davies, and V. Gabrielsen, 253–297. Oxford: Oxbow Books.

Riedel, A. 1994. "Archaeozoological investigations in North-Eastern Italy: The Exploitation of Animals since the Neolithic." *Preistoria Alpina* 30. 43–94.

Rombai, L. 2002. "Clima, suolo e ambiente." In *Storia dell'agricoltura italiana I. L'età antica. 1. Preistoria*, edited by G. Forni and A. Marcone, xvii–lxiv. Firenze: Edizioni Polistampa.

Roselaar, S. T. 2008. "Regional Variations in the Use of the *Ager Publicus*." In *People, Land and Politics: Demographic Developments and the Transformation of Roman Italy, 300 BC–AD 14*, edited by L. de Ligt and S. J. Northwood, 573–602. Leiden: Brill.

Roselaar, S. T. 2010. *Public Land in the Roman Republic: A Social and Economic History of Ager Publicus in Italy, 396–89 BC*. Oxford: Oxford University Press.

Roselaar, S. T. Forthcoming. "Republican Italy." In *A Companion to Ancient Agriculture*, edited by D. B. Hollander and T. Howe. Malden, MA: Wiley Blackwell.

Rosenstein, N. S. 2004. *Rome at War: Farms, Families, and Death in the Middle Republic*. Chapel Hill: University of North Carolina Press.

Rosenstein, N. S. 2008. "Aristocrats and Agriculture in the Middle and Late Republic." *Journal of Roman Studies* 98: 1–26.

Rosenstein, N. S. 2009. "Aristocrats and Agriculture in the Late Republic: The 'High Count'." In *Agricoltura e scambi nell'Italia tardo-repubblicana*, edited by J. Carlsen and E. Lo Cascio, 243–257. Bari: Edipuglia.

Rossiter, J. J. 1978. *Roman Farm Buildings in Italy*. Oxford: B.A.R.

Rossiter, J. J. 1981. "Wine and Oil Processing at Roman Farms in Italy." *Phoenix* 35: 345–361.

Sallares, R. 1995. "Molecular Archaeology and Ancient History." In *Food in Antiquity*, edited by J. Wilkins, D. Harvey, and M. Dobson, 87–100. Exeter: University of Exeter Press.

Sallares, R. 2007. "Ecology." In *The Cambridge Economic History of the Greco-Roman World*, edited by W. Scheidel, I. Morris, and R. P. Saller, 15–37. Cambridge: Cambridge University Press.

Saller, R. P. 2007. "Household and Gender." In *The Cambridge Economic History of the Greco-Roman World*, edited by W. Scheidel, I. Morris, and R. P. Saller, 87–112. Cambridge: Cambridge University Press.

Saller, R. P. 2012. "Human Capital and Economic Growth." In *The Cambridge Companion to the Roman Economy*, edited by W. Scheidel, 71–86. Cambridge: Cambridge University Press.

Santoro, S. 2017. "Crafts and Trade in Minor Settlements in North and Central Italy: Reflections on an Ongoing Research Project." In *The Economic Integration of Roman Italy: Rural Communities in a Globalizing World*, edited by T. de Haas and G. W. Tol, 263–295. Leiden: Brill.

Scarborough, J. 1969. *Roman Medicine*. Ithaca, NY: Cornell University Press.

Scheidel, W. 1994. "Grain Cultivation in the Villa Economy of Roman Italy." In *Landuse in the Roman Empire*, edited by J. Carlsen, P. Orsted, and J. E. Skydsgaard, 159–166. Rome: L'Erma di Bretschneider.

Scheidel, W. 1995. "The Most Silent Women of Greece and Rome: Rural Labour and Women's Life in the Ancient World, 1." *Greece & Rome* 42: 202–217.

Scheidel, W. 1996. "Finances, Figures and Fiction." *Classical Quarterly* 46: 222–238.

Scheidel, W. 1997. "Quantifying the Sources of Slaves in the Early Roman Empire." *Journal of Roman Studies* 87: 156–169.

Scheidel, W. 2004. "Human Mobility in Roman Italy, I: The Free Population." *Journal of Roman Studies* 94: 1–26.

Scheidel, W. 2005. "Human Mobility in Roman Italy, II: The Slave Population." *Journal of Roman Studies* 95: 64–79.

Scheidel, W. 2007a. "Demography." In *The Cambridge Economic History of the Greco-Roman World*, edited by W. Scheidel, I. Morris, and R. P. Saller, 38–86. Cambridge: Cambridge University Press.

Scheidel, W. 2007b. "Roman Funerary Commemoration and the Age at First Marriage." *Classical Philology* 102 (4): 389–402.

Scheidel, W. 2011. "The Roman Slave Supply." In *The Cambridge World History of Slavery*, edited by K. Bradley and P. Cartledge, 287–310. Cambridge: Cambridge University Press.

Scheidel, W., and S. J. Friesen. 2009. "The Size of the Economy and the Distribution of Income in the Roman Empire." *Journal of Roman Studies* 99: 61–91.

Scheidel, W., and S. Von Reden. 2002. *The Ancient Economy*. New York: Routledge.

Schwartz, M. 2016. "Bitumen, Greece and Rome." In *The Encyclopedia of Ancient History*, edited by R. S. Bagnall, K. Brodersen, C. B. Champion, A. Erskine, and S. R. Huebner. Malden, MA: Blackwell.

Sena Chiesa, G. 1985. *Angera romana. Scavi nella necropoli 1970–1979.* Rome: Giorgio Bretschneider Editore.

Serletis, A. 2001. *The Demand for Money: Theoretical and Empirical Approaches.* Boston: Kluwer Academic Publishers.

Shaw, B. D. 2013. *Bringing in the Sheaves: Economy and Metaphor in the Roman World.* Toronto: University of Toronto Press.

Skydsgaard, J. E. 1968. *Varro the Scholar: Studies in the First Book of Varro's De Re Rustica.* Copenhagen: Einar Munksgaard.

Small, A. M. 1974. "Excavations at Monte Irsi, Italy, 1971–1973." *Echos du Monde Classique/Classical Views* 18: 20–22.

Small, A. M. 2011. *Vagnari: il villaggio, l'artigianato, la proprietà imperiale.* Bari: Edipuglia.

Small, A. M., and R. J. Buck. 1977. "S. Giovanni di Ruoti, Southern Italy (1977)." *Echos du Monde Classique/Classical Views* 22: 5–8.

Small, A. M., S. G. Moncton, R. J. Buck, and C. J. Simpson. 1992. "Excavations at Gravina di Puglia, 1991: Interim Report." *Echos du Monde Classique/Classical Views* 36: 189–199.

Small, A. M., C. Small, R. Abdy, A. De Stefano, R. Giuliani, M. Henig, K. Johnson, P. Kenrick, T. Prowse, and H. Vanderleest. 2007. "Excavation in the Roman Cemetery at Vagnari, in the Territory of Gravina in Puglia, 2002." *Papers of the British School at Rome* 75: 123–229.

Solomon, J. 1995. "The Apician Sauce: *Ius Apicianum.*" In *Food in Antiquity,* edited by J. Wilkins, D. Harvey, and M. Dobson, 115–131. Exeter: University of Exeter Press.

Spurr, M. S. 1983. "The Cultivation of Millet in Roman Italy." *Papers of the British School at Rome* 51: 1–15.

Spurr, M. S. 1986a. "Agriculture and the *Georgics.*" *Greece and Rome* 33 (2): 164–187.

Spurr, M. S. 1986b. *Arable Cultivation in Roman Italy c. 200* BC–AD *100.* London: Society for the Promotion of Roman Studies.

Stek, T. D. 2010. *Cult Places and Cultural Change in Republican Italy: A Contextual Approach to Religious Aspects of Rural Society after the Roman Conquest.* Amsterdam: Amsterdam University Press.

Stern, E. M. 2008. "Glass Production." In *The Oxford Handbook of Engineering and Technology in the Classical World,* edited by J. P. Oleson, 520–547. Oxford: Oxford University Press.

Stewart, P. 2008. *The Social History of Roman Art.* Cambridge: Cambridge University Press.

Takaoğlu, T. Forthcoming. "Roman Anatolia." In *A Companion to Ancient Agriculture,* edited by D. B. Hollander and T. Howe. Malden, MA: Wiley Blackwell.

Tan, J. 2017. *Power and Public Finance at Rome, 264–49* BCE. New York: Oxford University Press.

Taylor, L. R. 1951. "Caesar's Agrarian Legislation and his Municipal Policy." In *Studies in Roman Economic and Social History in Honor of Allan Chester Johnson,* edited by P. R. Coleman-Norton, 68–78. Princeton: Princeton University Press.

Tchernia, A. 1983. "Italian Wine in Gaul at the End of the Republic." In *Trade in the Ancient Economy,* edited by P.D.A. Garnsey, K. Hopkins, and C. R. Whittaker, 87–104. Berkeley and Los Angeles: University of California Press.

Tchernia, A. 1986. *Le vin de l'Italie romaine: Essai d'histoire économique d'après les amphores.* Rome: École Française.

Temin, P. 2001. "A Market Economy in the Early Roman Empire." *Journal of Roman Studies* 91: 169–181.

Terrenato, N. 2012. "The Enigma of 'Catonian' Villas: The *De agri cultura* in the Context of Second-Century BC Italian Architecture." In *Roman Republican Villas: Architecture, Context, and Ideology,* edited by J. A. Becker and N. Terrenato, 69–93. Ann Arbor: University of Michigan Press.

Thaler, R. H. 2015. *Misbehaving: The Making of Behavioral Economics.* New York: W. W. Norton & Co.

Thibodeau, P. 2001. "The Old Man and His Gardens (Verg. *Georg.* 4.116–148)." *Materiali e discussioni per l'analisi dei testi classici* 47: 175–195.

Thibodeau, P. 2011. *Playing the Farmer: Representations of Rural Life in Vergil's Georgics.* Berkeley: University of California Press.

Thomas, R., and A. Wilson. 1994. "Water Supply for Roman Farms in Latium and South Etruria." *Papers of the British School at Rome* 62: 139–196.

Thompson, J. 1988. "Pastoralism and Transhumance in Roman Italy." In *Pastoral Economies in Classical Antiquity,* edited by C. R. Whittaker, 213–215. Cambridge: Cambridge Philological Society.

Thurmond, D. L. 2006. *A Handbook of Food Processing in Classical Rome: For Her Bounty No Winter.* Leiden: Brill.

Tol, G. W. 2017. "From Surface Find to Consumption Trend: A Ceramic Perspective on the Economic History of the Pontine Region (Lazio, Central Italy) in the Roman Period." In *The Economic Integration of Roman Italy: Rural Communities in a Globalizing World,* edited by T. de Haas and G. W. Tol, 367–387. Leiden: Brill.

Torrent, J. 2005. "Mediterranean Soils." In *Encyclopedia of Soils in the Environment,* edited by D. Hillel, 418–427. Amsterdam: Elsevier.

Toti, O. 1966. "Civitavecchia. Edificio rustico romano in località 'Monna Felice'." *Notizie degli scavi di antichità* 20: 79–90.

Toynbee, A. J. 1965. *Hannibal's Legacy: The Hannibalic War's Effect on Roman Life.* London: Oxford University Press.

Treggiari, S. 1969. *Roman Freedmen During the Late Republic.* Oxford: Oxford University Press.

Uden, J. 2010. "The Vanishing Gardens of Priapus." *Harvard Studies in Classical Philology* 105: 189–219.

Vaccaro, E., C. Capelli, and M. Ghisleni. 2017. "Italic Sigillata Production and Trade in Rural Central Italy: New Data from the Project 'Excavating the Roman Peasant'." In *The Economic Integration of Roman Italy: Rural Communities in a Globalizing World,* edited by T. de Haas and G. W. Tol, 231–262. Leiden: Brill.

Vaccaro, E., M. Ghisleni, A. Arnoldus-Huyzendveld, C. Grey, and K. Bowes. 2013. "Excavating the Roman Peasant II: Excavations at Case Nuove, Cinigiano (GR)." *Papers of the British School at Rome* 81: 129–179.

van Driel-Murray, C. 2008. "Tanning and Leather." In *The Oxford Handbook of Engineering and Technology in the Classical World,* edited by J. P. Oleson, 483–495. Oxford: Oxford University Press.

Veal, R. 2017. "Wood and Charcoal for Rome: Towards an Understanding of Ancient Regional Fuel Economics." In *The Economic Integration of Roman Italy: Rural Communities in a Globalizing World,* edited by T. de Haas and G. W. Tol, 388–406. Leiden: Brill.

Verboven, K. 2002. *The Economy of Friends: Economic Aspects of Amicitia and Patronage in the Late Republic.* Bruxelles Latomus.

Veyne, P. 1979. "Mythe et réalité de l'autarcie à Rome." *Revue des Études Anciennes* 81: 261–280.

von Stackelberg, K. 2013. "Columella." In *The Encyclopedia of Ancient History*, edited by R. S. Bagnall, K. Brodersen, C. B. Champion, A. Erskine, and S. R. Huebner, 1679–1680. Malden, MA: Blackwell.

Wallace-Hadrill, A. 1990. "Pliny the Elder and Man's Unnatural History." *Greece & Rome* 37: 80–96.

Wallace-Hadrill, A. 1998. "Horti and Hellenization." In *Horti romani: atti del Convegno internazionale, Roma, 4–6 maggio 1995,* edited by M. Cima and E. La Rocca, 2–12. Rome: L'Erma di Bretschneider.

Watson, A. 1998. *The Digest of Justinian.* Philadelphia: University of Pennsylvania Press.

White, K. D. 1967. *Agricultural Implements of the Roman World.* Cambridge: Cambridge University Press.

White, K. D. 1970. *Roman Farming.* Ithaca, NY: Cornell University Press.

White, K. D. 1973. "Roman Agricultural Writers I: Varro and his Predecessors." *Aufstieg und Niedergang der römischen Welt* 1(4): 439–497.

White, K. D. 1975. *Farm Equipment of the Roman World.* Cambridge: Cambridge University Press.

White, K. D. 1995. "Cereals, Bread and Milling in the Roman World." In *Food in Antiquity,* edited by J. Wilkins, D. Harvey, and M. Dobson, 38–43. Exeter: University of Exeter Press.

Whittaker, C. R. 1985. "Trade and Aristocracy in the Roman Empire." *Opus* 4: 49–75.

Wiedemann, T.E.J. 1981. *Greek and Roman Slavery: A Sourcebook.* Baltimore: Johns Hopkins University Press.

Wilkins, J. M., and S. Hill. 2006. *Food in the Ancient World.* Oxford: Blackwell.

Wilkins, J., D. Harvey, and M. Dobson. 1995. *Food in Antiquity.* Exeter: University of Exeter Press.

Wilson, A. 2008. "Villas, Horticulture and Irrigation Infrastructure in the Tiber Valley." In *Mercator Placidissimus: The Tiber Valley in Antiquity: New Research in the Upper and Middle River Valley: Rome, 27–28 February 2004,* edited by H. Patterson and F. Coarelli, 731–768. Rome: Quasar.

Wilson, A. 2009. "Approaches to Quantifying Roman Trade." In *Quantifying the Roman Economy: Methods and Problems,* edited by A. K. Bowman and A. Wilson, 213–249. Oxford: Oxford University Press.

Witcher, R. 2006. "Broken Pots and Meaningless Dots? Surveying the Rural Landscapes of Roman Italy." *Papers of the British School at Rome* 74: 39–72.

Witcher, R. 2016. "Agricultural Production in Roman Italy." In *A Companion to Roman Italy,* edited by A. E. Cooley, 459–482. Malden, MA: Wiley Blackwell.

Witcher, R. 2017. "The Global Roman Countryside: Connectivity and Community." In *The Economic Integration of Roman Italy: Rural Communities in a Globalizing World,* edited by T. de Haas and G. W. Tol, 28–50. Leiden: Brill.

Woolf, G. 1990. "Food, Poverty and Patronage: The Significance of the Epigraphy of the Roman Alimentary Schemes in Early Imperial Italy." *Papers of the British School at Rome* 58: 197–228.

Yntema, D. 1993. "Greeks, Natives and Farmsteads in South-Eastern Italy." In *De Agricultura: In Memoriam Pieter Willem De Neeve,* edited by H. Sancisi-Weerdenburg, R. J. van der Spek, H. C. Teitler, and H. T. Wallinga, 78–97. Amsterdam: J. C. Gieben.

Zanda, E. 2011. *Fighting Hydra-Like Luxury: Sumptuary Regulation in the Roman Republic.* London: Bristol Classical Press.

Zanker, P. 1998. *Pompeii: Public and Private Life.* Cambridge, MA: Harvard University Press.

Zohary, D., M. Hopf, and E. Weiss. 2012. *Domestication of Plants in the Old World: The Origin and Spread of Domesticated Plants in South-West Asia, Europe, and the Mediterranean Basin.* Oxford: Oxford University Press.

Index

acne 31
acorns 30, 32
ager publicus 57, 90n6, 99, 103n7
agrestes 31, 48; *see also* peasants
Alexandria 13, 14, 85
alimenta 93, 102, 104n58
almonds 31–32
amicitia 91n36
amphorae 5, 10, 47, 51, 72, 101
amurca 29
Anatolia 19n89
animal feed 25, 30
annona 71
antidotes 31
Antium 101
Antonine Plague 101, 102
Antoninus Pius 13
Antonius, M. 6
Apicius 80n56
apiculture 32, 33, 68–71
apples 7, 29
Apulia 103n10, 104n37
aqueducts 21, 90
Aricia 90n6
artisans 8, 49, 78, 83, 84
Asia Minor 99
asparagus 32, 44, 74
astrologers 52, 60n55, 84
Athens 5, 70
Augustus 13, 14, 33, 67, 71, 97, 101–102
autarky 1–3
aviaries 31, 65, 66, 69
axes 10, 42, 44

bakers 71, 104n36
bandages 31
bankers 1, 41, 78
bark 31, 48, 70
barley 24–26, 31

barrels 46, 72, 101
basil 44, 51
baskets 5, 28, 31, 42, 44, 48, 50, 70
baths, bathing 8, 67, 73, 90, 94
beans 29–31
bee-glue 70
beekeeping 6, 30, 69
beets 32, 74
behavioral economics 3
billhooks 44
birds 6, 7, 31, 65–66, 69
bitter vetch 29–30
bitumen 50
blacksmith 41, 49
Bodel, John 4–5, 37n38
books 30
Boscoreale 10, 59n19
Boscotrecase 12
Bowes, Kim 11, 58n5
braziers 45
bread 23–26
bread wheat 23 24
brine 50
Brunt, Peter 6
buckets 42, 45
butchers 64–65, 67, 91n15
butter 68

cabbage 4, 32, 39n119, 74
Caesar, C. Iulius 6, 99, 100
Calabria 103n10
Cales 42, 47
Calpurnius Siculus 9, 67
Campania 21, 42, 75, 99, 104n37
canals 13
candles 31
capital 65, 72, 87, 102
capital, human 23
Capua 40, 42, 47, 75

carob 29
carrots 32
carts 42, 48
Casinum 42
cassia 55
Cassius 14
caterpillars 27, 70
Catiline 82n104, 99
Cato the Elder 4–5, 52, 62
Cato the Younger 104n35
cattle 34, 46, 57, 64–5, 100
celery 32
cement 42
centuriation 11
cereal 15, 23–27, 99
Ceres 51, 52
chaff 26–27
chains 42, 45
charcoal 74–75
cheese 50, 64, 67–68, 78
cherry trees 29
chestnuts 31–32
chickens 34, 65, 66
chickpeas 29–30
children 23, 68, 77, 84, 102
Chile 82n107
Cicero, M. Tullius 12, 40, 54, 57, 62
Cincinnatus 12, 14
cinnamon 34
circitores 91n24
cisterns 8, 27, 41, 42
citron 29
Claudius 101
clients 51, 62, 77, 83, 84, 88, 89
climate 20–22, 34
Clodius, P. 104n35
clothing 34–5, 47, 67, 51, 75, 95
coinage 10, 41, 53–7, 63, 98, 100
colonies 42, 57, 90, 93, 98, 100
Columella 7, 9–10, 27–8, 102
combs 30, 45, 46
competition 49, 65, 74, 85, 86, 98
confiscation 15, 88, 89, 93, 98, 99, 100
contracts 29, 56, 60n41, 87, 88
convertible husbandry 39n124, 64, 65
Cosa 10
cosmetics 33, 69
control marks 13
cows 12, 60n49, 67, 68, 80n46
credit 53, 56, 77–78, 86, 89, 96
cremation 55
crop-selection 38n83
cucumbers 32
Cumae 31

curses 51
cuttings 27, 29, 72
cypresses 88

de Ligt, Luuk 58n4, 89
debt 53, 85, 89, 99, 101
debt slavery 89
deforestation 21
demography 22–23, 97
deodorants 30
digging 14, 28, 77
Diocletian 80n41
Diodorus Siculus 45
disease 24, 34, 64
doctors 41, 49, 54–55, 67
dogs 7, 12, 34, 46, 86
dolia 28, 47, 49, 50
Domitian 101
donkeys 7, 34, 43, 46, 50, 64
Doody, Aude 9
dormice 32, 35, 65, 100
drainage 21
drought 15, 21, 24, 26, 29, 32, 87
ducks 65, 66
Duncan-Jones, Richard 17n21, 57
durum 23–24
dyes 31
Dyson, Stephen 9

eggs 34, 66, 78
Egypt 13, 14, 21, 24, 50, 87, 101
einkorn 25
Elba 45
elms 74, 88
emmer 23–26
endowments 54
Epirus 6, 48, 97
equinox, fall 39n119
equinox, spring 33
Erdkamp, Paul 2, 60n39, 71, 73, 76, 89
ergastula 14, 17n41
erosion 22
esparto grass 42, 48
Etna 21
Etruria 7, 97, 99
Evans, J. K. 3

faber, fabri 42, 45
fabric 30, 34
fairs 85
Falerii 69
fall 24–28, 30, 34, 68, 70, 76
fallow 26
family 7, 15, 23, 34, 41, 56–57, 84

fans 46
Feltria 54
fences 74
fenugreek 29, 30
Fidenae 91n35
Field Museum 10
field survey 11, 84, 97
figs 29, 74, 82n102
fineware 11, 95
firewood 74, 78, 87, 94
fish 6, 7, 9, 35, 65
fish, salted 51, 68
fishermen 12
fishing 13, 31
fishmongers 91n15
fishponds 13, 65
flatulence 31
flax 26, 30–31, 45, 48, 75
flaxseed oil 31
flour 24, 25, 30, 48
flowers 33, 54, 73, 74, 78, 95
fodder 25–6, 29–31, 34–5, 71, 74, 77
forests 21, 32, 62, 65
Forum Holitorium 74
fowling 13, 31
foxes 27
Foxhall, Lin 28, 29, 64
frankincense 34, 57
Frayn, Joan 2, 68, 79n17, 103n21
Frederiksen, Martin 3
Fregellae 79n29, 90n6
frescoes 12–14, 67
friends 4, 15, 51, 56, 77–78, 83, 87–9
Fronto 73
fruit 26, 29, 64, 74, 98
fruit trees 7, 29, 64, 74
frumentationes 71, 89, 96, 102n3, 104n35
Fucine lake 102n3
fuel 28, 35, 68, 71, 73, 74–5
fullers 1, 49, 50, 66
funerals 14, 55–57, 69
furniture 51, 94

Gades 7
Gallia Belgica 8
Gallia Narbonensis 8
game 9, 19n91, 39n130, 64, 65
gardens 9, 10, 32–3, 73–4, 94
garlic 26, 32
garum 51, 68, 95
geese 25, 65–66
gifts 16n7, 51, 82n106, 87, 89
glass 51, 57, 64, 95
glue 64, 70

gluten 24, 26
goats 7, 12, 34, 46, 64–65, 67–68
Goodchild, Helen 21
gourds 33, 74
Gracchi, the 99
Gracchus, Tiberius Sempronius 57, 97, 98
grafting 27, 29, 32
granaries 11, 23, 47
grape syrup 72
grapes 12, 13, 27–28, 44, 48, 72
Greece 38n98
groundwater 21
gum 50
gypsum 30, 49

ham 47, 50, 79n17
Hannibal 4, 98
harbors 13, 104n46
harness 42, 48
Harris, William V. 75
harvest, harvesting 13–14, 22, 44, 48–9
hatchets 44
hay 30, 44, 71, 77
hayforks 10
hazelnuts 31–32, 34
headaches 31
Healy, John 8
Heinrich, Frits 26
helmets 43
hemp 26, 31, 48
hens 25
herbs 33
Herculaneum 67, 79n34
hides 34, 64, 65, 71; *see also* skins
Hin, Saskia 21
Hispania Tarraconensis 8
hoes 10, 43–44
Holleran, Claire 3, 64, 74
honey 31, 33, 68–71, 78, 87, 95
Hopkins, Keith 2
Horace 82n106, 88, 89
Horden, Peregrine 3, 21, 36n1
horses 7, 34, 59n30
horticulture 9, 12, 32
hunting 13, 31
huts 31, 48

ignorance 36n32, 77
Illyricum 82n104
in-kind payments 48, 52, 68, 86, 87, 98
incense 52, 55
inheritance 23, 42, 54, 57, 84, 98

insects 29, 46; *see also* pests
intercropping 29
interest 28, 62, 66
interest rates 53, 62, 78, 99, 102
irrigation 20, 21, 26, 27, 33, 63
Isola Sacra 19n91

Jashemski, Wilhelmina 10
javelins 8, 43
Jongman, Willem 15, 22, 53–54, 66, 72
juniper 34

Kelsey Museum of Archaeology 10
keys 42, 46
knives 42, 44, 57, 71
Koppen, Vladimir 20
Kron, Geoffrey 9, 10, 64, 65, 68, 74
Kronenberg, Leah 6

labor 49, 76–7, 96
laborers, hired 40, 49, 77, 88, 96
ladders 49
ladles 49
lampreys 35
land, marginal 22, 33, 34, 64
landlords 15, 77, 87
Latium 7, 10
Launaro, Alessandro 11, 49, 90n8,
    103n10
Leach, Eleanor 14
lead 47, 50
legumes 15, 29–30, 74, 78, 86
lentils 26, 29–30, 31
leporaria 65
lettuce 32, 74
ley farming 39n124
life expectancy 23
lime 49, 75
linen 30–31, 75
Ling, Roger 12, 14
Lipkin, Sanna 34
livestock 33–35, 41, 46, 64–5, 68, 94
Livy 50, 57
loans 52, 63, 77–8, 89, 96, 98, 102
locatio conductio 87
locks 42, 46
looms 42, 47
Lucania 42
Lucretius 9, 47
lumber 74
Luni 101
lupins 29–30, 47, 74
luxury 5, 66–7, 74, 86, 98

Maecianus 55
magic 58n1
Maiuro, Marco 101, 103n10
malaria 22
manure, manuring 7, 26, 34, 43, 44, 71
marble 51
Marcomanni 100
Marcone, Arnaldo 7, 16n11
maritime villa 18n84, 59n35
Marius, C. 13, 99
markets 2–4, 40–1, 62–3, 71–4, 83–90,
    93–8
marls 60n44
marriage 23, 84, 88
marshes 13, 34, 45
Marzano, Annalisa 10, 18n71, 59n35,
    60n44
mats 31, 49
Mattingly, David 28–9, 73
mattocks 42, 43
meat 34, 50, 64–6, 68, 71, 100
medicine 31, 33, 34, 52, 54, 63, 70
melons 32
merchants 41, 45, 63, 72, 78, 85–6, 89
Methana 81n94
mice 27
microclimates 35, 36n1
milk 67–8, 78, 95
millets 26
millstones 8, 50
mint 44
Minturnae 42, 47, 97
Misenum 8, 69
mola salsa 52
money 15, 53–5, 57, 62–3
moneylending 62, 63, 64, 77–8, 99
Monte Testaccio 28
Moretum 9, 32, 47, 74
Morley, Neville 74, 85
moths 29, 70
mules 7, 34, 43, 46, 58n1
mulsum 69
muria 51
mushrooms 33
must 28
myrrh 34, 55

nails, iron 57, 58n10
Naples, Bay of 59n25
neighbors 21, 52–3, 78, 86, 88–9
Nero 8, 32, 57, 67
nets 31
nitrogen 30

North Africa 8, 13, 22, 29, 74, 101
Novum Comum 7
Nuceria 91n35
nundinae 2, 40–1, 51, 52, 85, 86, 88
nuts 31–2, 33, 74, 78

oaks 34, 88
oats 25–6
obaerarii 82n104
Octavian 100
olive oil 12, 28–9, 46–7, 49,
    73, 101
olive trees 28–9, 44
olives 28–9, 50, 71, 73, 76
onions 32
opium 33
Oplontis 12
ornithones 65–6
osiers 28
Ostia 50, 69, 85, 94
oxen 31, 34, 48, 50, 57, 75

Pakistan 82n107
paleosols 18n75
Parentalia 52, 54, 73
parsley 32
parsnip 32
partnerships 35, 75, 89
pastio villatica 6, 65, 98, 100
pastures 34, 35, 44, 62, 100
paterfamilias 40, 47, 51, 74, 76, 84
patronage 87–8
patrons 15, 51, 77, 84, 89, 96
peafowl 65, 66
peas 26, 29, 30
peasants 1–2, 43, 52, 74, 76, 89, 95; *see
    also rustici*
peculia 81n100, 92n43
perfume 14, 33, 51, 55, 69, 95
Pergamum 57
pests 24, 27–8, 34, 46, 64; *see also* insects
Petronius 3, 70, 102n5
pigeons 66
pigs 7, 34, 64, 65, 68
pines 50, 88
pisciculture 35, 65, 78
piscinae 65
pistachios 31–2
pitch 50, 62, 86, 94, 95
pitchforks 46
Pliny the Elder 7–10
Pliny the Younger 8, 55, 87, 89, 94,
    101, 102

plowing 7, 12–15, 44, 64
plows 43
plowshares 43, 45
plums 29
Plutarch 4, 62, 75
Po river 26
poison 30, 69, 70
pollen 11
polyculture 27, 93
Pompeii 10, 12, 30, 42, 48, 87
Pompey 6, 69
Poppaea 57, 67
poppies 33
porridge 24, 25
precipitation 20–1, 26; *see also* rainfall
preservation, food 7, 26, 29, 33, 50,
    69, 95
presses, pressing 28–9, 41, 42, 47–8,
    49, 68
props 27–8, 31, 40, 74
proscription 6, 15, 90, 99, 100
protein 24, 29
pruning 5, 27, 28, 44
puls 24
pulses *see* legumes
Purcell, Nicholas 3, 21
Puteoli 45

Quadi 100
quarries 75
quern 11, 48

rainfall 26; *see also* precipitation
raisins 27, 72
rakes 43, 46
Rathbone, Dominic 11, 76
rationality 2, 3
reaping 13, 22, 26, 46
Reay, Brendon 5
reciprocity 84, 87–9, 93, 96
redistribution 15, 57, 89–90, 93, 96,
    98, 99
rennet 68
rent, remission of 91n30
resin 49
retting 30
rice 26
risk 2, 27, 51, 78, 85, 87, 100
rivers 13, 21, 22
roads 90, 98
Roman Peasant Project 10, 39n124
Rome 42, 68, 72–5, 97–102
ropes 4, 31, 42, 47–8

Rosenstein, Nathan 26, 27, 57, 58, 72, 98–9
rush, rushes 31, 48
rustici 7, 29, 44; *see also* peasants
rye 25

sacrifice 42, 51, 52, 54, 64
sails 30, 81n92
Sallares, Robert 20, 24
Saller, Richard 16n1, 23, 104n37
salt 47, 50–1, 68, 86, 94, 95
Samnium 57
sand 58n8, 75
saws 44
scrolls 51
scythes 44
sea-tortoises 34
Second Punic War 4, 40, 97
security 4, 15, 53, 85, 100
self-sufficiency 1–4, 35, 57, 66, 93
Seneca 41
sesame 33
Settefinestre 10
sharecropping 42, 87, 102
Shaw, Brent 22
shearing 34–5
shears 34, 44
sheds 42
sheep 34, 64–8, 71
shepherds 12–13, 67–8, 84, 96
shipwrecks 10
shoes 47
shovels 44
Sicily 13, 57, 62, 97, 101
sickles 13, 19n90, 43, 44
Simulus 9, 32, 47, 74
skin 67, 69
skins 34, 46, 65, 72; *see also* hides
slave revolts 97
slaves 14, 23, 47, 48, 84, 89, 101
smallholders 35, 41, 48–9, 65, 76–77, 86–9, 96
snakebites 32
soap 31
soil 20–1, 43–4
soldiers 4, 5, 12, 25, 65, 67, 69
solstice, winter 33
sowing 7, 24–6, 30, 31, 33, 43, 51
spades 42, 44, 46
Spain 13, 29, 101
sparrows 26
Spartacus 99
spelt 25, 37n41
spinning 30–1, 66
sponges 49

spring 26, 30, 31, 33, 34, 70
springs 13, 21, 62
Spurr, Steven 4, 26
squab 66
Stabiae 18n88
stables 42
Stek, Tesse 51
stockbreeding 33
storage 46–7, 75
storms 15, 21, 85
Strabo 77
Strabo, C. Julius Caesar 99
strainers, straining 5, 31, 42, 67
strawberries 74
Studius 13
subsistence 35, 68, 84, 95
Suessa 42
Sulla 99, 100
sulphur 49
summer 20, 76
sumptuary legislation 65, 98, 100
suovetaurilia 51
superstition 44, 52
swords 43

tablets 70
tanning 64
Tarentum 79n29, 101
Tarracina 32
taxes, taxation 89, 97
Temin, Peter 85
tenancy 87
tenants 84, 87, 89, 94–6, 100, 101
terra sigillata 59n36
Terrenato, Nicola 5
textiles 47, 50, 66–7, 94; *see also* clothes
thieves, theft 81n102, 84
Thibodeau, Philiip 9
Third Punic War 4
threshing 7, 13, 24–5, 44, 46
threshing floor 19n91, 48
thrushes 82n102
Thurmond, David 9, 51, 72
Tiberius 14, 32, 101
Tibullus 14
tiles 42, 47
timber 46, 53, 88
Titus 8
toads 26
torches 31
traders 71, 74, 83, 84, 86
Trajan 13, 101
transhumance 34, 68
transportation 3, 20, 26–7, 72–3, 85–7
trellises 27

tributum 89
Trimalchio 3, 70, 94
tubers 32
turnips 26, 32
Tuscany 10, 75, 104n37

Ulpian 54, 55, 56
Umbria 10, 68
underemployment 60n39, 76–77

Vagnari 57
Varro 6, 9, 49
Veal, Robyn 75
vegetables 32–33, 73–4, 86, 95, 98
vehicles 42, 48
Venafrum 42
Vespasian 8
Vesuvius 8, 10, 101
vetch 29, 30
veterans 46, 63, 90, 98, 100, 101
Veyne, Paul 3
Via Sacra 70, 74
vilica 7, 40, 47, 52
vilici 7, 40, 45, 52, 84, 89
Villa Arianna 18n88
Villa Farnesina 12
Villa Publica 6
villages 84, 86, 89
villas 10, 13, 51, 90
vinitor 28
viticulture 27–8, 58, 72–3, 76, 101

Vitruvius 13
von Stackelberg, Katharine 7

wagons 48, 75
Wallace-Hadrill, Andrew 9
walnuts, walnut trees 31–2
warfare 87, 99, 100
water supply 7, 20, 32, 35, 42
wax 49, 70–1
weather 20, 21, 24, 26–7, 62, 71, 76
weaving 3, 49, 66
weeds, weeding 19n91, 26, 27, 29, 43, 44
wells 42
wheat 13, 23–7
White, K. D. 4, 6, 7, 43, 44, 68
wine 27–8, 46–7, 69, 72–3, 101
wine consumption 57, 72
winnowing 7, 19n91, 24, 30, 48
winter 20, 29, 45, 70–1
Witcher, Robert 11, 22, 102
woods 13, 34, 94
wool 34–5, 66–7
worms 26

Xenophon 51

yarn 34
yields 24, 25, 26, 27, 29, 65, 71, 100
yokes 13, 42, 48

Zanker, Paul 14